Springer Series in Materials Science

Volume 258

The Springer Series in Materials Science covers the complete spectrum of materials physics, including fundamental principles, physical properties, materials theory and design. Recognizing the increasing importance of materials science in future device technologies, the book titles in this series reflect the state-of-the-art in understanding and controlling the structure and properties of all important classes of materials.

More information about this series at http://www.springer.com/series/856

John O. Milewski

Additive Manufacturing of Metals

From Fundamental Technology to Rocket Nozzles, Medical Implants, and Custom Jewelry

 Springer

John O. Milewski
Los Alamos National Laboratory (Retired)
Santa Fe, NM
USA

ISSN 0933-033X   ISSN 2196-2812   (electronic)
Springer Series in Materials Science
ISBN 978-3-319-86348-1   ISBN 978-3-319-58205-4   (eBook)
DOI 10.1007/978-3-319-58205-4

© Springer International Publishing AG 2017
Softcover reprint of the hardcover 1st edition 2017
This work is subject to copyright. All rights are reserved by the Publisher, whether the whole or part
of the material is concerned, specifically the rights of translation, reprinting, reuse of illustrations,
recitation, broadcasting, reproduction on microfilms or in any other physical way, and transmission
or information storage and retrieval, electronic adaptation, computer software, or by similar or dissimilar
methodology now known or hereafter developed.
The use of general descriptive names, registered names, trademarks, service marks, etc. in this
publication does not imply, even in the absence of a specific statement, that such names are exempt from
the relevant protective laws and regulations and therefore free for general use.
The publisher, the authors and the editors are safe to assume that the advice and information in this
book are believed to be true and accurate at the date of publication. Neither the publisher nor the
authors or the editors give a warranty, express or implied, with respect to the material contained herein or
for any errors or omissions that may have been made. The publisher remains neutral with regard to
jurisdictional claims in published maps and institutional affiliations.

Printed on acid-free paper

This Springer imprint is published by Springer Nature
The registered company is Springer International Publishing AG
The registered company address is: Gewerbestrasse 11, 6330 Cham, Switzerland

Preface

The exciting new field of 3D printing has captured the imagination of makers and artists envisioning Star Trek type replicators, organic free-form designs and the desktop fabrication of everything from food and toys to robots and drones. A natural extension of this desire is to capture thoughts and dreams using metal due to its strength, durability and permanence.

Today, 3D printing and the field of additive manufacturing (AM) have received a lot of attention due to the introduction of personal 3D printers for the home, multi-million dollar government funding of additive and advanced manufacturing programs and corporate investments in research and development centers. Wild enthusiasm has been created within some sectors of industry and finance and most importantly among young people, creating the possibility of a rewarding career in additive manufacturing. The best way to temper this enthusiasm without losing momentum is to offer a balanced view of where the technology is today and where it can be tomorrow. As makers, how do we prepare for this opportunity? As a business owner, how can this affect my bottom line or that of my competitors? How mature is the technology and what long-term strategic advantages might it hold? A discussion of AM accomplishments and challenges, without all the hype, is needed to instruct, motivate, and create a devoted group of followers, learners, and new leaders, to fuel the passion and create the future of this technology.

This book is an introductory guide and provides learning pathways to 3D metal printing (3DMP) of near net shaped, solid free-form objects. That is objects that require little finishing to use and do not rely on design constrained by the limitations of current fabrication methods. Additive manufacturing is a term that broadens the scope of 3D metal printing to include a wide range of processes that start with a 3D computer model, incorporate an additive fabrication process, and end up with a functional metal part. The distinction between 3DMP and AM processes is blurring due to the rapid evolution of the many competing methods used to make a 3D metal part. In this book we will use both 3DMP and AM references, with a preference to AM.

The book presents a comprehensive overview of the fundamental elements and processes used to "3D print" metal. The structure of the book provides a roadmap of where to start, what to learn, how it all fits together, and how additive manufacturing can empower you to think beyond conventional metal processing to capture your ideas in metal. In addition, case studies, recent examples, and technology applications are provided to reveal current applications and future potential. This book shows how affordable access to 3D solid modeling software and high-quality 3D printing services can enable you to ascend the learning curve and explore how 3D metal printing can be put to work for you. This method of access enables us to begin our learning without the need to invest in the high-cost of professional engineering software or commercial additive manufacturing machines.

Those processes that sinter a bed of metal powder, fuse powder, or wire using high energy beams, or those hybrid processes that combine both additive and subtractive manufacturing (SM) methods may all fall under the category of AM. AM related processes have evolved at a dizzying pace spawning an avalanche of acronyms and terms, not to mention new companies being born, acquired and left behind.

In this book I will attempt to be internally consistent with the terminology and generic enough in their descriptions to minimize reliance on company names and trademarks. Rather than put a trademark symbol after every occurrence of a trademarked name, I will use names in an editorial fashion only and to the benefit of the trademark owner, with no intention of infringement of the trademark nor endorsement of the company. I admire the efforts of all these companies, past, present, and future and hope they succeed in these early days of technology development and adoption.

Together, we will look ahead and offer predictions of how 3DMP and AM will integrate into society and the global economy of a smaller, flatter world. We will combine our thoughts and dreams, amplified by advances in computing and information technology, to think of better ways to harness and transform metal into objects that serve us and help create an enduring future.

High-cost commercial additive manufacturing machines can range in price from hundreds of thousands of dollars to millions of dollars, but this does not mean we cannot begin to explore the technology without one. The price is sure to drop to levels affordable to small- and mid-size businesses, with metal printing services following suit. Hybrid versions of 3D plastic printing technology and low-cost versions of 3D weld deposition systems for the hobbyist are already in development. The current momentum of innovation in this rapidly changing field will provide affordable access to high quality 3D metal printers by the time we are ready to use them. In some cases we are already there. In this spirit, we proceed by adopting the analogy of you, the maker, as a *hitchhiker* and commercial 3D metal printers as the *vehicles* needed to manufacture your parts and solidify your dreams.

To achieve this goal, the book begins by providing the reader with a foundational understanding of how to learn and apply fused metal deposition to 3D printed parts. To understand keywords, phrases, technical terms, and concepts, you need to understand and speak the language of AM. These terms and jargon are listed and defined, within the context of AM processing and are located in the Glossary at the end of this book.

It is hard to separate the hype from the fact using common Web searches to discern amateur from professional opinion. Web searches using Google Scholar[1] provide links to a rich body of technical papers and published works, and in some cases provide open access to technical publications. Peer-reviewed technical publications are available for purchase although persons new to the field often need to establish a broader foundation of knowledge to fully benefit from the latest reported research.

Industry reports such as the Wohlers Report,[2] considered by many to be the bible of 3D printing, present a yearly update to the latest developments within the technology but do not provide the technical detail of how these processes work. This book directs the readers to articles in online publications and magazines, covering the additive manufacturing industry, to provide in-depth coverage of technical advancements.

In this book we strive to provide cost effective references, search terms in italics, Web links and references to complement a consistent technical description of AM metal printing processes, allowing readers to engage in just-in-time learning as directed by knowledgeable and appropriate sources.

Additive manufacturing (AM) refers to a large and complex field encompassing model-based design engineering, computer-aided design (CAD) and computer-aided manufacturing (CAM) software, process engineering and control, materials science and engineering and industrial practice. To date, a comprehensive "how to" text on the AM processes of metals, more commonly referred to as 3D metal printing, is not yet available. Technical experts working in AM often have expertise in one or more of these fields but few have a deep understanding of the entire technical spectrum. Publications are spread across a very wide range of journals and Web based sources. The issue is, books specifically focused on "How to 3D print with metal" do not exist. This need for a single source of entry-level information provides another motivation for this book.

Those with a strong interest in additive manufacturing technology often do not know where to start to get a high-level structured view of the processes as applied to metals. This may be intimidating or confusing to beginners and those considering "dabbling" in or exploring the technology. You need not be a student, a maker, a metal fabricator, or a business owner to see the potential in AM or have an interest in how to 3D print metal. AM is complex enough that those embarking on the path of "experiential self-learning" are often stymied by lack of preparation or basic knowledge needed to succeed in those first few projects required to assess the

---

[1]Google Scholar provides access to a wide range of technical papers, citations and patents. http://scholar.google.com/, Setting Google or Google Scholar alerts is a good way to stay up to date on the latest developments of the technology and market place.

[2]Wohlers, T., & Caffrey, T. (2014), Wohlers Report 2014—3D Printing and Additive Manufacturing State of the Industry, Wohlers Associates, http://www.wohlersassociates.com/, (accessed March 30, 2015).

technology and gain confidence. Most books on "How to 3Dprint" are popular books on 3D printing plastics, some are overhyped or strictly forward-looking. Additive manufacturing textbooks often attempt to cover the entire spectrum of materials and sacrificing important design considerations, process details, or application considerations as applied to metal. "How to" books on AM are good start, but if your interest is in metals you should find a book that focuses on these AM materials.

Vendor-supplied operation manuals or Web links to recommend "standard conditions" exist for specific materials using a specific system, but the truth is most owners and users of high end commercial systems are also engaged in trial and error development, otherwise known as learning the hard way. Vendor-supplied guide-lines are either very generic or strictly prescriptive, imparting a recipe but without an in-depth understanding of why we do what we do. Vendors often protect standard operating parameters as proprietary, keeping them secret from the machine owners, also obscuring the workings of the technology. Much has been written in the technical literature regarding *how to* 3D print metals, but more often than not there is little mention of how *not to* 3D print with metals, or the information is presented as a partial work leaving out relevant details. Knowing what can go wrong is often just as important as knowing how to get it right.

What is 3D metal printing and how does it differ from 3D printing with plastics or other materials? How can I create complex metal objects and move beyond the constraints of conventional metal processing? How can I learn the basics, explore and choose the 3D metal printing process that is right for me? In this book you will learn you do not need a degree in engineering, or a million dollar 3D metal printer, to reach the cutting-edge of additive manufacturing.

An additional goal of this book is to help you decide what you need to get started, what types of software, materials, and processes are right for you, what additional knowledge is required, and where to get it. For those just starting out or those embarking on a new career path, AM holds promise to be a good profession, offering a rewarding, well-paying career from the production floor, to the corporate research and development (R&D) lab, to a viable commercial business opportunity. Emerging careers in additive and advanced manufacturing are hot real estate and if you have the will, there is surely a way. If this book inspires you to take either path, we have succeeded twice, as some of this book will be sure to remain with you on your journey.

I begin by emphasizing the fundamental understanding of 3D metal printing, identifying the building blocks, why we do what we do, and what is important to you the maker. The book provides information often overlooked related to critical applications, such as those in aerospace, automotive, or medical fields and the rigorous path to certification. The average maker may never build a rocket ship or reach for the stars or design and build a unique medical device that saves lives, but you never know. This book will introduce the reader these topics and applications.

3D additive manufacturing moves us toward a more complex and information-rich environment. We are not just creating the "soul of a new machine," we are creating its DNA as well. This product DNA information generated and stored along the way will include a cradle-to-grave documentary of design, fabrication, and service life. Not only do we create the DNA, we grow the object and put it to work.

Santa Fe, NM, USA   John O. Milewski

Acknowledgements

I would like to thank the members of the AWS D20 committee, the ANSI/America Makes AMSC, ASTM F42, and EWI AMC who have shared their technical experience and valuable insights into AM metal technology.

In addition I would like to thank Matt Johnson, Van Baehr, Jim Crain, Dan Schatzman, Ben Zolyomi, Jim and Linda Threadgill, John Hornick, and Bill Stellwag for their views on the book content and scope.

I acknowledge the contributions of all the other researchers, entrepreneurs, and makers in the field, mentioned in the book only in passing and some not at all. If not, be assured I applaud your efforts as well and wish you luck creating and catching this new wave of technology.

Santa Fe, NM, USA

John O. Milewski

Contents

About the Author

John O. Milewski received his B.S. in Computer Engineering from the University of New Mexico and his M.S. in Electrical Engineering from Vanderbilt University. He began his technical career with 5 years of metal fabrication experience ranging from heavy industry production as an ASME code welder to light manufacturing and applied research. He spent 32 years at Los Alamos National Laboratory, in positions including Welding Technologist, Engineer, Team Leader, Experimental Component Fabrication Program Manager, and Group Leader for Manufacturing Capability. He is currently retired from the Lab, writing and consulting as APEX3D LLC regarding the new and exciting applications of AM technology.

His technical expertise includes arc systems electron beam, laser welding, robotics, sensing and controls, and the joining of less common metals. His work experience also includes CAD/CAM/CNC model-based engineering, process modeling, and simulation with validation methods to include residual stress measurement. In addition, he served 2 years in the late 1980s as Vice President of Synthemet Corporation, an entrepreneurial high tech start-up, with the goal of development and commercialization of 3D additive manufacturing of metals.

He is author and co-author of numerous publications related to high energy beam processing and process modeling. His awards include an R&D 100 Award for Directed Light Fabrication, Fellow of the American Welding Society (AWS) and the AWS Robert L. Peaslee Award. He is inventor or co-inventor for a number of patents related to laser welding and additive manufacturing.

He has had extensive formal collaborations with universities and sponsored students resulting in refereed publication and patenting. His professional society involvement included Chairman, Co-Chairman, and advisor of AWS committees related to High Energy Beam, Electron Beam, and Laser Beam Welding. In addition, he currently serves as an advisor to the AWS D20 Additive Manufacturing committee and provides peer review to technical publications of AWS and ASM International.

His international technical contributions include US Delegate to the High Energy Beam Welding Commission of the IIW (International Institute of Welding), Invited Keynote Lecturer at the 58th Annual Assembly and International Conference of IIW, and the AWS R.D. Thomas Award winner for his international contributions and committee work on the harmonization of international standards.

Acronyms

3DFEF	3D Finite Element Fabrication
3DMP	3D Metal Printing
3DP	3D Printing
AI	Artificial Intelligence
AM	Additive Manufacturing
AMF	Additive Manufacturing File Format
B2B	Business to Business
CAD	Computer Aided Design
CAE	Computer-Aided Engineering
CAM	Computer-Aided Manufacturing
CFD	Computational Fluid Dynamics
CMM	Coordinate Measurement Machine
CNC	Computerized Numerical Control
CSG	Computed Solid Geometry
CT	Computed Tomography
DED	Directed Energy Deposition
DED-EB	Directed Energy Deposition Electron Beam (see EB-DED)
DED-L	Directed Energy Deposition Laser (see L-DED)
DED-PA	Directed Energy Deposition Plasma Arc (see PA-DED)
DLD	Direct Laser Deposition
DMCA	Digital Millennium Copyright Act
DMD	Direct Metal Deposition
DMLS	Direct Metal Laser Sintering
DRM	Digital Rights Management
DTRM	Discreet Transfer Radiation Model
DTSA	Defend Trade Secrets Act
EB	Electron Beam
EB-DED	Electron Beam Directed Energy Deposition (see DED-EB)
EBAM	Electron Beam Additive Manufacturing
EBF3	Electron Beam Free Form Fabrication

EBM	Electron Beam Melting
EB-PBF	Electron Bean Powder Bed Fusion (see PBF-EB)
EBSM	Electron Beam Selective Melting
EBW	Electron Beam Welding
ECM	Electro-Chemical Milling
EDM	Electrode Discharge Machining
ELI	Extra Low Interstitial
ES&H	Environment, Safety & Health
F2F	Factory to Factory
FDM	Fused Deposition Modeling
FEA	Finite Element Analysis
FEF	Finite Element Fabrication
FoF	Factory of the Future
FZ	Fusion Zone, in welding
GFR	Geometric Feature Representation
GMA	Gas Metal Arc
GMAW	Gas Metal Arc Welding
GTA	Gas Tungsten Arc
GTAW	Gas Tungsten Arc Welding
HAZ	Heat Affected Zone, in welding
HCF	High Cycle Fatigue
HDH	Hydride DeHydride
HEPA	High-efficiency Particulate Arrestance
HIP	Hot Isostatic Press
HT	Heat Treatment
IGES	Initial Graphics Exchange Specification
IoT	Internet of Things
IP	Intellectual Property
IR	Infrared Radiation
ISRU	In Situ Resource Utilization
IT	Information technology
IV&V	Independent Verification and Validation
LB	Laser Beam
LBW	Laser Beam Welding
LCF	Low Cycle Fatigue
LDT	Laser Deposition Technology, RPM Innovations
LENS	Laser Engineered Net Shape, Optomec
LMD	Laser Metal Deposition
L-PBF	Laser Powder Bed Fusion (see PBF-L)
M2M	Machine to Machine
MAST	Math, Science and Technology
MBE	Model Based Engineering
MEMS	Micro-Electro-Mechanical Systems
MIG	Machine Inert Gas (welding)
MRO	Maintenance, Repair and Overhaul

MSDS	Material Safety Data Sheet
NDT	Non-Destructive Testing
NEMS	Nano-Electro-Mechanical Systems
NURBS	Non-Uniform Rational B-spline
OIM	Orientation Imaging Microscopy
OM	Optical Microscopy
PA	Plasma Arc
PA-DED	Plasma Arc Directed Energy Deposition (see DED-PA)
PAW	Plasma Arc Welding
PBF	Powder Bed Fusion
PBF-EB	Powder Bed Fusion Electron Beam (see EB-PBF)
PBF-L	Powder Bed Fusion Laser (see L-PBF)
PDM	Product Data Management
PLM	Product Lifecycle Management
PM	Powder Metallurgy
ppb	Parts per billion
PPE	Personal Protective Equipment
ppm	Parts per million
PREP	Plasma rotating electrode process
PSD	Particle Size Distribution
PT	Penetrant Testing
QA	Quality Assurance
RPD	Rapid Plasma Deposition
RT	Radiographic Testing
SEM	Scanning Electron Microscopy
SLA	Stereo Lithography
SLM	Selected Laser Melting
SLS	Selective Laser Sintering
SM	Subtractive Manufacturing
STEM	Science Technology Engineering and Math
STEP	Standard for the Exchange of Product model data
STL	Standard Tessellation Language
TEM	Transmission Electron Microscopy
TIG	Tungsten Inert Gas (welding)
TPM	Technological Protection Measures
TRL	Technology Readiness Level
UAM	Ultrasonic Additive Manufacturing
UC	Ultrasonic Consolidation
UT	Ultrasonic Testing
UTS	Ultimate tensile strength
UV	Ultraviolet radiation
VR	Virtual Reality
WAAM	Wire + Arc Additive Manufacturing
XRF	Energy dispersive X-Ray Fluorescence

AM Road Map and Hitchhiker's Guide

In this book you will learn how the power of computer based solid models and the advent of 3D printing services can enable you, the maker, to create complex metal objects beyond the constraints of conventional metal processing.

Chapter 1 sets the stage and takes us from the dawn of metal processing to the dawn of 3D metal printing. But where are we now? Where do you want to go? When do you take the first step?

In Chap. 2 we take a whirlwind tour of the 3D metal printing and additive manufacturing landscape. To whet your appetite for the journey, we show you novel designs and applications made possible by AM metal. What application fields are hot? How do these applications take us beyond conventional metal processing and point us toward the future?

Chapter 3 asks the question: What is in your backpack? How are you positioned to take advantage of this exciting new technology? We provide a high-level overview of the type of AM processes, examples of AM users and identify the drivers of AM technology adoption.

In Chap. 4 we learn to speak the language of metal, what is it? What properties of metal are relevant to building 3D printed objects? What metals work best for 3D printing and why? How do you select which metal is best for you?

Chapter 5 is the next building block of your AM metal knowledge foundation. Understanding high energy heat sources is fundamental to understanding how metal melts, fuses, and then cools into a solid part. Should you use a laser or an electron beam? What is the difference between metal sintering and metal fusing? When is using a plasma arc or gas metal arc source the best option? What are the pros and cons of each? This chapter will provide a basic understanding of the heat sources used in 3D metal printing and which one is best for fabricating your components.

Computers, 3D models, computer motion systems, and controls are fundamental subsystems of all 3D printers. Chapter 6 will tell you how and why 3D metal printing systems use computerized models and controls, and how these systems differ when using metals from those used for 3D printing plastics and polymers.

Chapter 7 will show there is much to be learned from the technologies foundational to AM, such as 3D plastic printing, laser weld cladding, and powder metallurgy. Maintaining strong links between AM and these foundational technologies will continue to foster understanding and new ideas beneficial to all.

AM metal printers come in all shapes and sizes, ranging from multi-million dollar machines capable of depositing jet fighter backbones to "3D shape welders" based on homemade motion systems and arc welding equipment. The user needs to understand the details of how each method starts with a model and ends up with a part. Depending on the material and application, the end product may be substantially different. What are these differences and why should you care? What skills are required along the way, from generating a model to creating a finished part? Chapter 8 will describe current AM system configurations in detail to equip the user to choose the right process and services based upon the requirements of the design and end use of the part.

For those with experience in design for conventional metal fabrication, Chap. 9 will define a new design space for thinking "outside the box". We will contrast design space thinking for 3D printed plastics versus that needed to build a 3D metal part. We will build upon the existing body of conventional metal working knowledge, complement it, transform it, and in some cases take it to levels not possible a few short decades ago. We will also introduce design concepts for hybrid processes that combine conventional materials processing with additive materials and shapes.

A common misperception is that AM provides the ability to pull the part out of the 3D printer, bolt it up and drive off or fly away with it. This is rarely the case. Chapters 10 and 11 will provide the knowledge of how to develop the process, the pre- and postprocessing operations and the critical considerations for choosing design features, materials, process conditions, and parameters. Unlike plastics, post processing AM deposited metals may include highly specialized equipment and costly post processing operations. Knowledge of these operations is needed to achieve full functional performance of your metal part. In addition, these chapters will help you to make informed decisions regarding whether to outsource AM fabrication to a service provider or to purchase and develop and an in-house capability.

In Chap. 12 we step back and take a broader view to survey trends in government, industry, universities, and business systems to plot the course where the technology will have the greatest impact in the next ten years. We consider global trends, the increasing role of information and how AM may connect it to our world, our environment, and ultimately to ourselves. Our dreams need not be constrained by time, space, or money, but they will inevitably be connected to one another. Finally we will leave you with our view of the destination of AM technology into the future.

Chapter 1
Envision

Abstract Metal working has played a key role in the development of human civilization. Naturally occurring metals such as gold, began to take form in manmade metal objects as the Stone Age transitioned from gathered objects into the earliest forms of metal processing technology. Metal objects took the form of personal adornment, symbols of power, and tools of conquest. Thousands of years passed as metal extraction and forming technology slowly advanced to include alloys such as bronze and metals such as iron. The age of discovery led to the identification, extraction, refinement, and use of new metallic elements, alloys, and manufacturing processes. The dawn of the computer age enabled significant technical advances in decades rather than centuries. Information available to all leads to the convergence of technologies empowering individuals to design and manufacturing complex metal objects. That which was once the providence of Kings and Pharaohs, empires, armies, and captains of industry is now within our reach. This chapter provides a brief introduction to technical milestones leading to the development of additive metal process technology. Metal has empowered mankind with the ability to capture our visions, realize our dreams, and build objects that extend the power of our thoughts in time and space. Metal is often hidden in nature and requires time, labor, and energy to extract and form into useful objects. This elusive and mysterious nature of metal has been part of the allure in its ownership as the expense, capabilities, and skills needed to create complex metal objects has most often been out of reach to all but a few. Metal and energy, harnessed by man's dreams, grant us the power to capture the present and create the future. Epics of human progress have been defined by metals such as bronze, iron, and steel; energized by the sun, fire, and electricity. Gold and silver have built world power and adorned our most prized possessions. Steel has built armies, skyscrapers, bridges, railways, and pipelines. Copper has joined together voices of people from across the world. But, the world is changing. In this new century, thoughts are created, captured, and shared as data across the planet at the speed of light. Information on any subject is at our fingertips, where and when it is needed. But words and images are not enough, we still need things to have, hold, and use. In a world where information is called the new power and bitcoins emerge as a new currency, it can all disappear in a flash. Metal endures. But first, where did it all start?

J.O. Milewski, *Additive Manufacturing of Metals*, Springer Series
in Materials Science 258, DOI 10.1007/978-3-319-58205-4_1

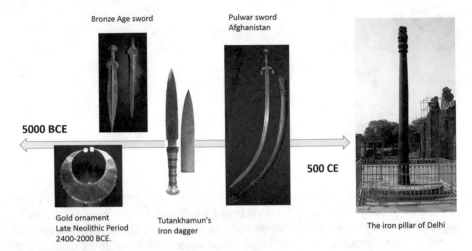

Fig. 1.1 Metal working in ancient times. Gold ornament[1], Bronze Age sword[2], Tutankhamun's Iron Dagger[3], Pulwar sword Afghanistan[4], The iron pillar of Delhi[5]

1.1 Evolution of Metalworking

- In Neolithic times, 6000–3000 BCE, natural forms of metal such as gold and copper are cold worked into items such as fetishes, talismans, or personal adornment (Fig. 1.1).
- The Bronze Age begins, 3000 BCE, copper is alloyed with arsenic or tin and is used in castings and forgings to make strong tools and weapons. Iron is found in objects such as daggers for the Pharaohs of Egypt.
- The Iron Age begins, 1400–1200 BCE, the Hittites and others begin to learn how to forge iron tools and weapons for conquest.
- Crucible steels, such as Wootz in India (∼300 BCE), are developed for weaponry and later Damascus steel weapons are renowned for strength and

---

[1]"Gold lunala from Blessington, Ireland, Late Neolithic/, Early Bronze Age, c. 2400 BC–2000 BC, Classical group," Johnbod, CC-SA-3.0, https://en.wikipedia.org/wiki/Gold_lunula#/media/File: Blessingon_lunulaDSCF6555.jpg.

[2]Bronze Age Sword, Apa-Schwerter aus Rumänien, Dbachmann, licensed under CC by SA 3.0. https://commons.wikimedia.org/wiki/File:Apa_Schwerter.jpg.

[3]Comelli, D., D'orazio, M., Folco, L., El-Halwagy, M., Frizzi, T., Alberti, R., Capogrosso, V., Elnaggar, A., Hassan, H., Nevin, A., Porcelli, F., Rashed, M. G. and Valentini, G. (2016), The meteoritic origin of Tutankhamun's iron dagger blade. Meteoritics & Planetary Science, 51: 1301–1309. 10.1111/maps.12664, copyright John Wiley and Sons.

[4]Afganistan Pulwar Sword, © Worldantiques, Own Work, is licensed under CC by SA 3.0, https://commons.wikimedia.org/wiki/File:Afghanistan_pulwar_sword.jpg.

[5]The iron pillar of Delhi, Photograph taken by Mark A. Wilson (Department of Geology, The College of Wooster), public domain, https://commons.wikimedia.org/wiki/File:QtubIronPillar.JPG.

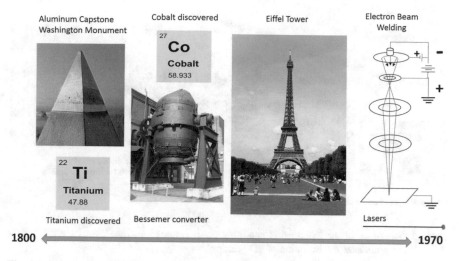

Fig. 1.2 Discovery and mass production of metals, Aluminum Capstone[6], Bessemer converter[7], Eiffel Tower, the invention of electron beam welding and laser[8]

durability. Structures, such as the Iron Pillar of Delhi, $\sim$400 CE, remain to this day as testament to the skill of ancient metallurgists and craftsman.

- Discoveries of new metals occur in the mid-1700s to the late 1800s AD. Important metals such as titanium, tungsten, cobalt, and aluminum, are discovered. New extraction processes from raw ore are developed. Aluminum objects are made for kings and czars. A capstone pyramid, made of the rare metal aluminum, is placed atop the Washington Monument for is dedication in 1885 (Fig. 1.2).
- The Bessemer process for the industrial-scale production of steel is developed in the mid-1850s.
- The dream of technical genius, engineer, and metallurgist Gustav Eiffel is completed in time for the 1889 World's Fair in Paris. The Eiffel Tower was constructed using 2.5 million rivets.
- The carbon arc process for welding metal is developed in the late 1800s.

---

[6]Washington monument aluminum capstone, public domain, http://loc.gov/pictures/resource/thc. 5a48088/, Reproduction Number: LC-H824-T-M04-045 Library of Congress, Washington, D.C., USA.

[7]The copyright by Dave Pickersgill, licensed for reuse under CC-SA 2.0 l, https://upload. wikimedia.org/wikipedia/commons/thumb/6/64/Bessemer_Convertor_-_geograph.org.uk_-_ 892582.jpg/450px-Bessemer_Convertor_-_geograph.org.uk_-_892582.jpg.

[8]Eiffel Tower, seen from the Champ de Mars, Paris, Franc, © Waithamai—Own work, is licensed under CC-SA 3.0, https://commons.wikimedia.org/wiki/File:Eiffel_Tower_Paris_01.JPG.

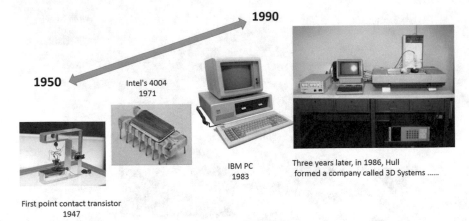

Fig. 1.3 The dawn of the information age. First point contact transistor[9], Intel's 4004[10], IBM PC[11], Three years later the invention of 3D printing[12]

- In the 1910s–1950s, military applications such as building ships and aircraft, push the development of weld processing of steel, aluminum and titanium alloys. Gas tungsten arc welding and gas metal arc welding wee developed in the 1940s.
- Electron beam welding is invented in the late 1950s. Lasers are first demonstrated in 1960. Space and nuclear applications push the development of refractory and reactive metal joining for metals such as tantalum, niobium, and zirconium.
- CNC lathes, using punched tape, revolutionize the machining industry in the 1960s and 1970s. The SR-71 Reconnaissance Aircraft, with a structure made of 85% titanium alloy, flies its first mission in 1968. Designer Clarence "Kelly" Johnston was responsible for many of the design's innovative concepts.

1.2 Advent of Computers

- Early computers for controlling machines are developed in the 1950s (Fig. 1.3).
- Microprocessor chip-based computers are developed in the 1970s.

---

[9]https://commons.wikimedia.org/w/index.php?curid=24483832, Courtesy of Unitronic under CC BY-SA 3.0: https://creativecommons.org/licenses/by-sa/3.0/.

[10]"Intel's 4004," https://commons.wikimedia.org/w/index.php?curid=3338895, Courtesy of firm Intel under CC BY-SA 3.0: https://creativecommons.org/licenses/by-sa/3.0/.

[11]"IBM PC," https://commons.wikimedia.org/w/index.php?curid=9561543.4, Courtesy of Rubin de Rijcke, under CC BY-SA 3.0.

[12]Courtesy of 3D Systems, reproduced with permission.

- The January 1975 issue of <u>Popular Electronics</u> features the Altair 8800 computer on its cover. Ed Roberts, president and chief engineer at MITS, Albuquerque, NM, creates a kit computer that helps launch a tech revolution.[13]
- Desktop personal computers see widespread usage in the 1980s. Microprocessor-based machine controllers evolve in sophistication.
- Computer graphics and 3D computer-aided design (CAD) software evolves with hardware in the 70s, 80s and 90s. Feature-based parameterized solid models see adoption in aerospace, automotive and other high-end manufacturing industries.
- Computer-aided model-based engineering and networks tie together the design process with fabrication processes from forming and machining to inspection in the 1980s–2000s.

1.3 Invention of 3D Printing

- Stereolithography using ultraviolet (UV) lasers to cure photopolymers into 3D shapes is developed in 1984, by Chuck Hull, founder of 3D Systems Corporation. The technology utilizes machine commands, derived from slicing a computer solid model to direct the production of 3D shapes using polymers. Other early additive processes used to make solid free-form objects for visualizing form and testing fit are developed and described later in the book.
- Development of AM processes for metal is developed in the 1980s and 1990s. Research at universities, national labs, and industrial R&D Labs result in technical collaborations.

Additive manufacturing brings the power to create complex, free-form metal objects to anyone with a computer, access to a 3D model and an AM printing service. Empowered by information technology, AM bypasses many of the costly steps, equipment, and skills of metal working, allowing solid free form designs to be transformed into near net-shaped metal objects with the click of a key. Recent media attention and the advent of the 3D printing of a wide range of materials has created a hype and in some cases a false expectation, that in the near future any material will be rapidly printed into any size and shape. This is not yet the case and in many instances may never be, but the technology is moving in that direction. Large corporate investments are being made, in a very dynamic market both in the production of metal printing machines but also in the demonstration and adoption of the technology.

---

[13]The Kit That Launched the Tech Revolution, Forrest M. Mims, Make: Vol. 42, Dec 2014/Jan 2015.

1.4 Key Take Away Points

- Early metal working evolved over thousands of years beginning with naturally occurring metals such as gold for jewelry and evolving to simple alloys such as bronze for weapons, iron for tools and ultimately steel.
- The age of discovery identified many new metallic elements and developing processes used for mass production. The industrial age saw a rapid increase in the metal working progress made during a few hundred years.
- The information age kicked in with the advent of computers and microelectronics. Access to digital information and control evolved rapidly over a few decades and now touches nearly all aspects of technology.
- 3D printing, rapid prototyping and, additive manufacturing have developed at the intersection of information and material processing technology and are now being adopted at a dynamic pace with significant developments emerging every year.

Chapter 2
Additive Manufacturing Metal, the Art of the Possible

Abstract Novel applications and designs showcase the power and potential of 3D printing and additive manufacturing of metal. This chapter identifies the market segments where AM is making inroads and having the greatest impact. The momentum in advances of AM metal technology, historically driven by rapid prototyping for engineering applications, is rapidly changing and moving toward the production of high value components made of advanced materials. Application examples include critical products such as those certified for use in aerospace and for medical hardware. In addition, customized artistic designs and one of a kind personalized items are being created on demand. Unique designs and functions are being incorporated into parts unthought-of only yesterday. The potential to transform industry and realize significant cost and energy savings is demonstrated by the use complex cooling channels in hardware for the tool and die industry to high-performance heat exchangers and wear resistant coatings in the energy, oil and gas industries. Repair and remanufacturing of components with every increasing complexity using multiple and advanced materials is being demonstrated. This chapter provides examples of components and products within the hottest segments of AM metal technology, introducing the art of the possible.

2.1 AM Destinations: Novel Applications and Designs

So, where we are today? What is the state of development and application of AM technology as applied to metals? What is unique to these applications that make them attractive to using AM? (Fig. 2.1) While the best of the technology is most likely under wraps in the back shop of a corporate research lab or cutting-edge fabrication shop, there are a number of applications we would like to showcase. Some of these examples are technology demonstrations, forward looking marketing

© Springer International Publishing AG 2017
J.O. Milewski, *Additive Manufacturing of Metals*, Springer Series in Materials Science 258, DOI 10.1007/978-3-319-58205-4_2

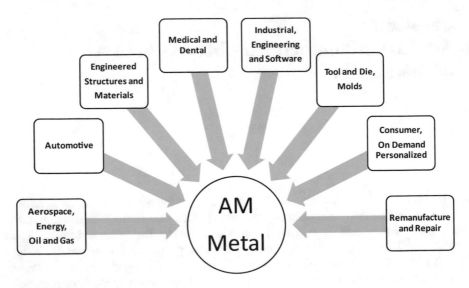

Fig. 2.1 Applications of AM metal

examples, or honest to goodness functional prototypes, but they all serve the purpose of pushing pins and drawing circles on the AM roadmap.

A number of examples of out of the box thinking and novel designs are provided as vectors for inspiration or perhaps outright destinations. What is a killer application and how does one get there? Truly unique applications are emerging every day. Some are destined to become product lines and quiet money makers, others may serve as mental launch pads for inventions and a method beyond today's thinking.

Figure 2.2 is perhaps the most widely publicized AM part, the GE Aero LEAP fuel nozzle, featuring a cobalt chrome alloy and other materials. It combines 18 components into one part with complex passageways and cutting-edge design offering higher durability and efficiency. With 19 nozzles per engine and a future production rate of 1700 engines per year, GE Aviation has set the goal of 32,000 nozzles per year when in full production, 100,000 parts by 2020. GE has invested $3.5 billion dollar in new plants to produce these nozzles. These nozzles are already in flight testing.

In another example General Electric has completed testing its Advanced Turboprop (ATP) technology demonstration engine which will power the all-new Cessna Denali single-engine aircraft. The engine is 35%-additive manufactured

Fig. 2.2 General
Electric LEAP nozzle[1]

featuring a *clean sheet* design used to validate additive parts, reduce the weight by 5% while contributing a 1% improvement in specific fuel consumption (SFC).[2] In another additive test program, the CT7-2E1 demonstrator engine was designed, built and tested in 18 months, reducing more than 900 subtractive manufactured parts to 16 additive manufactured parts.

2.2   Artistic

Artistic applications of 3D metal printing are leading the way toward exploration of entirely new designs, shapes and processes. Some of these capture the essence of freeform, emotional design. A design and part by Bathshiba Sculpture LLC[3] is one example, shown in Fig. 2.3.

---

[1]Courtesy of GE Aviation, reproduced with permission.

[2]GE Aviation press release, October 31, 2016, GE tests additive manufactured demonstrator engine for Advanced Turboprop, http://www.geaviation.com/press/business_general/bus_20161031a.html, (accessed January 20, 2017).

[3]Bathshiba Sculpture LLC Web site, http://bathsheba.com/, (accessed April 6, 2015).

Fig. 2.3 3D printed metal
sculpture[4]

As software and material become cheaper, artistic access to solid free form
design tools will allow a further expansion into the world of emotional design. As
music, color, video, and other forms of dynamic audio and visual 2D art can evoke
emotional response or inspirational experience, so will 3D virtual reality
(VR) headsets and the 3D VR experience. Capturing moments of 3D VR and
bringing them back to the physical world will be enabled by AM. This will include
kinetic artwork and parts that change in time within the local environment of use.

3D printing machines designed for precious metals[5] feature a powder management process developed for the jeweler and watchmaking industries, ensuring full
accountability of the valuable powders and providing quick metal changeover
through a cartridge-based system. The 3D metal printing machines used to create
jewelry can be smaller and relatively less expensive than machines used to print
automotive and aerospace parts. Artwork and jewelry does not require the same
levels of certification and control needed by aerospace, automotive and medical
devices, therefore making jewelry an attractive market for additive processing.
Artistic designs that cannot be produced in metal by any other method are made
possible while using less material and streamlining the production of custom
made-to-order pieces. Hollow structures with internal supports allow the fabrication
of larger pieces with the desired strength but without the weight or cost or a solid
piece. AM systems such as shown in Fig. 2.4a feature small build volumes ideal for
the rapid fabrication of small pieces such as jewelry while minimizing the total
volume of precious metal powder stock. They use a small laser focal spot sizes,
providing excellent detailed resolution, allowing the creation of fine features and
structures, as shown in Fig. 2.4b and c.

---

[4]Courtesy of Bathsheba Grossman, reproduced with permission.
[5]Cooksongold website, http://www.cooksongold-emanufacturing.com/products-precious-m080.
php, (accessed August 13, 2015).

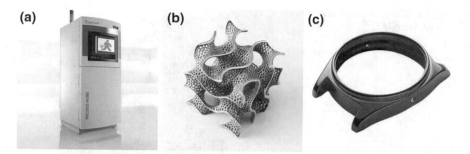

(a) **(b)** **(c)**

Fig. 2.4 **a** Direct Metal Laser Sintering machine for jewelry. "M 080 Direct Precious Metal 3D Printing System,"[6] **b** Sculptural design printed in gold.[7] **c** 3D printed gold watchcase.[8]

2.3 Personalized

Renishaw and Empire Cycle teamed up to build the first design of the titanium bicycle as described in an article from Engineering and Technology magazine, "First 3D printed bike enters record books," by Alex Kalinauckas.[9] Figure 2.5 shows the frame components as-fabricated in sections within the AM machine build volume. Figure 2.6 shows the assembled frame with wheels and additional bicycle components. Technology demonstrations such as this highlight the ability to produce personalized designs out of specialty and lightweight materials such as titanium. Complex shapes with lightweight internal strengthening structures and flowing organic forms allow the combination of engineering and artistic features to produce unique one of a kind individualized objects.

3D scanning and printing is now commonly used in the fabrication of custom-fit hearing aids and other such personal devices. Although currently made in polymers, the hearing aid example shows the potential for 3D scanning and printing to disrupt market places and radically change products made specifically for you. Mass produced items have the appeal of low cost but in some cases the benefit of a customized item made specifically for you will offer the greatest value. As scanning and digital definition of our bodies becomes common place, every human-to-object interface

---

[6]Courtesy of Cooksongold and EOS, reproduced with permission.

[7]Manufactured and designed by Cooksongold, reproduced with permission.

[8]Manufactured by Cooksongold, designed by Bathsheba Grossman, reproduced with permission

[9]Article from Engineering and Technology magazine, "First 3D printed bike enters record books", March 18, 2015, by Alex Kalinauckas, http://eandt.theiet.org/news/2015/mar/3d-bikeframe.cfm, (accessed March 26, 2015).

Fig. 2.5 Titanium bike frame
as-built using AM[10]

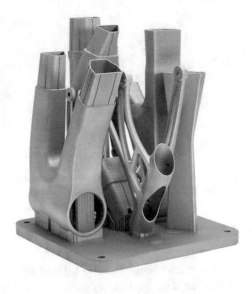

Fig. 2.6 Titanium bike frame
as assembled[11]

holds the potential for customization. As an example, a mobile app[12] may be used to
order a personalized piece of jewelry. Personalized rings with the initials of a loved
one, can be printed in various precious metals as shown in Fig. 2.7. Custom made
and personalized items, such as golf club heads,[13] are being produced by Ping.
Although out of the price range for many, these types of items can infer personal
taste and passion for the sport, as well as status, for all-out equipment freaks
(Fig. 2.8). Any sport, personal item or household fixture with a high end market can
be a target of innovative and unique designs made possible using AM metal.

---

[10]Courtesy of Renishaw, reproduced with permission.

[11]Courtesy of Renishaw, reproduced with permission.

[12]Love by me website and app, http://love.by.me/, (accessed August 13, 2015).

[13]3D article, http://3dprint.com/46036/golf-equipment-manufacturer-ping-introduces-golfs-first-
3d-printed-putter/, (accessed August 13, 2015).

Fig. 2.7 Personalized jewelry[14]

Fig. 2.8 Custom golf club head[15]

2.4 Medical

"Disruptive" applications for AM are beginning to emerge into the manufacturing mainstream. One such application is that of dental devices, where small custom-fit crowns and dental implants are disrupting the historic methods for the fabrication of these components. Figure 2.9 shows dental crowns and bridges produced by direct

---

[14]Courtesy of Skimlab and Jweel, reproduced with permission.
[15]Courtesy of Ping, reproduced with permission.

Fig. 2.9 Additively manufacturing dental hardware[17]

metal laser sintering (DMLS). In one example[16] an EOS M 100 DMLS machine fuses Cobalt Chrome SP2 alloy, a medical material using a certified and qualified process. Small lot size, high precision and high value products such as these are seeing wide adoption.

Another application soon to be widely realized is metal medical implants, as certifications for medical use are being approved for human use in the European Union (EU) and US. Over 50,000 medical devices have been implanted for the medical industry as produced by the electron beam melting (EBM) additive manufacturing process alone.[18] The benefits provided by AM are those of rapid production of personalized fit items for direct use, such as for implants, or secondary uses, such as drill guides and fixtures using the patient's own medical imaging to create 3D models anatomically matched devices. The accuracy of direct AM parts is sufficient for these applications, while the surface finish or porous structures offer advantages for bone ingrowth. These complex engineered surfaces are cleaned and sterilized offering a biological fixation intended to replace cemented fixation to optimize the implant–host interface. Figure 2.10 shows a 3D printed titanium cranial implant on a 3D printed skull model.

In another example, Stryker has received 510(k) clearance from the U.S. Food and Drug Administration today for its *Tritanium* PL Posterior Lumbar Cage, spinal

---

[16]EOS application for dental crowns, http://www.eos.info/eos_at_ids_additive-manufacturing, (accessed August 13, 2015).

[17]Courtesy of EOS, reproduced with permission.

[18]Arcam White Paper, Optimizing EBM Alloy 718 Material for Aerospace Components, Francisco Medina, Brian Baughman, Don Godfrey, Nanu Menon, downloaded from, http://www.arcam.com/company/resources/white-papers/, (accessed January 20, 2017).

Fig. 2.10 Titanium skull implant[19]

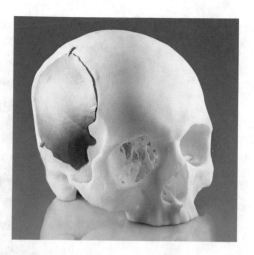

implant device for patients with degenerative disk disease.[20] The device is manu-factured via a 3D additive manufacturing process using their proprietary Tritanium technology, a novel highly porous titanium material designed for bone ingrowth and biologic fixation.

The FDA[21] Website for Medical Application of 3D printing provides additional information as well as links to draft guidance on the Technical Considerations for Additive Manufactured Devices to obtain public feedback. When finalized and in effect, the guidance will advise manufacturers who are developing and producing devices through 3D printing techniques with recommendations for device design, manufacturing, and testing.

Materials offering sufficient strength and biocompatibility are those currently used in medical devices, such as cobalt chrome and titanium alloys, which are easily fabricated using AM. Specialty metals such as tantalum may also see wider use in AM produced devices or AM deposited surfaces. Such medical devices command a high price and fit well within the build volume of powder bed fusion processes.

---

[19]Courtesy of 3T RPD, reproduced with permission.

[20]Stryker press release, http://www.stryker.com/en-us/corporate/AboutUs/Newsroom/Product Bulletins/169618, (accessed January 20, 2017).

[21]FDA Web site, Medical Applications of 3D Printing, http://www.fda.gov/MedicalDevices/ProductsandMedicalProcedures/3DPrintingofMedicalDevices/ucm500539.htm, (accessed January 20, 2017).

Fig. 2.11 Titanium
propulsion tank[23]

Courtesy of Lockheed Martin

The titanium propulsion tank shown here is 16″ in diameter.
Subsequent parts could be as large as 50″ in diameter.

2.5 Aerospace

Lockheed Martin and Sciaky have demonstrated the use of AM for the creation of a
titanium propulsion tanks using EBAM, as shown in Fig. 2.11. In this case, the
EBAM process is used to create a rough blank shape that can later be machined into
a shape that would otherwise need to be formed by obtaining commercially
available titanium plate, pressing into shape, then machining. Pressing would
require a forming punch and die and a large hydraulic press. Vessels of various
sizes would require a costly punch and die for each shape.[22]

---

[22]Sciaky press release detailing the example in Fig. 2.11, http://www.sciaky.com/news_and_
events.html, (accessed March 26, 2015).

[23]Courtesy of Lockheed Martin, reproduced with permission.

Figure 2.12 show a full-scale rocket engine part 3D printed out of copper by NASA.[24] The additively manufactured part is designed to operate at extreme temperatures and pressures and demonstrates one of the advanced technologies NASA is evaluating for use in fabricating parts for the mission to the planet Mars. In another application, Aerojet Rocketdyne has fabricated and demonstrated the hot-fire testing of a rocket engine thrust chamber made using AM deposition of a copper alloy.[25] Figure 2.13 shows a liquid oxygen/gaseous hydrogen rocket injector assembly, built using additive manufacturing technology, being hot-fire tested at NASA Glenn Research Center. The potential reduction in fabrication lead times and costs provides strong motivation for evaluating the AM technology. Space and aerospace applications require strict procedures and certification for processes and components. Significant saving may be realized in the reduction of the number of certified parts and processes, such as joining, used to produce a component. The reduction in weight can result in significant savings during the launch into space escaping the gravity well of earth or fuel saving during commercial aircraft flights. The reduction in material waste during fabrication of expensive specialty materials such as nickel-based alloys or titanium is also an important factor in justifying the use of additive manufacturing. A panel at the "Technology Development and Trends in Propulsion and Energy", 2015 AIAA Propulsion and Energy Forum[26] describes the benefit of designing hardware with complex shapes and features not possible using conventional methods. Additional benefits in system efficiency and environmental factors such as noise and emissions may also be realized.

In a business case study, the Airbus Group EADS Innovations performed an eco-assessment analysis as applied to a standard Airbus A320 nacelle hinge bracket, shown in Fig. 2.14 and strove to include detailed aspects of the overall lifecycle: from the supplier of the raw powder metal, to the equipment manufacturer EOS, to the end-user, Airbus Group Innovations. An entire lifetime assessment contrasted costs and savings of each method along the entire manufacturing chain from cradle

---

[24]NASA 3D Prints the World's First Full-Scale Copper Rocket Engine Part, Tracy McMahan, April 21, 2015, http://www.nasa.gov/marshall/news/nasa-3-D-prints-first-full-scale-copper-rocket-engine-part.html, (accessed May 15, 2016).

[25]NASA and Aerojet Rocketdyne Successfully Tests Thrust Chamber Assembly Using Copper Alloy Additive Manufacturing Technology using copper alloy additive manufacturing technology, http://globenewswire.com/news-release/2015/03/16/715514/10124872/en/Aerojet-Rocketdyne-Hot-Fire-Tests-Additive-Manufactured-Components-for-the-AR1-Engine-to-Maintain-2019-Delivery.html#sthash.UU5Yuc9e.dpuf, (accessed May 14, 2016).

[26]"Technology Development and Trends in Propulsion and Energy," a panel at the 2015 AIAA Propulsion and Energy Forum. http://www.aiaa-propulsionenergy.org/Notebook.aspx?id=29179, (accessed August 13, 2015).

Fig. 2.12 Copper rocket
nozzle[27]

Fig. 2.13 Testing of an
additive manufactured rocket
nozzle[28]

to grave, indicating a lifetime cost saving primarily due to reduced weight (titanium
versus steel, lightweight design). Future comparisons with a wider scope of options,
such as epoxy composites, to determine the cost of environmental effects will shed
additional light on potential costs or benefits of all AM/SM options.

[27]Courtesy of NASA.
[28]Courtesy of NASA Glenn Research Center.

Fig. 2.14 A design for
Additive Manufacturing
meeting lightweight targets[29]

EOS and Airbus Group Innovations Team[30] (now the EADS Innovation Works) cites another study associated with the weight reduction benefits of AM designs with respect to energy consumption and the reduction of CO_2 emissions by nearly 40% over the full lifecycle of a conventionally cast steel aircraft bracket in comparison to an additive manufactured direct metal laser sintered titanium bracket with optimized topology. A savings of 25% in the reduction of titanium scrap and a possible weight savings of 10 kg per aircraft was also cited.

2.6   Automotive

Formula 1 race car design teams are benefitting from the design freedom and rapid prototype/testing to speed fabrication cycles to gain competitive advantage off the track. In these cases, cost is a secondary consideration while weight reduction and design freedom are paramount. These sorts of critical application components, made of plastics, metals, or composites are not subject to the same testing and certification constraints as commercial man-rated components, thus providing a high-performance test bed for these components. As they say, racing improves the breed and this applies to the materials, designs, methods, and machines here as well. Such applications provide a proving ground for AM technologies, although the success stories and detailed methods will be tightly held as company confidential information. A steering knuckle part for a race car fabricated by DMLS is shown in Fig. 2.15. Two additional examples include a light twin-walled drive shaft and

---

[29]Courtesy of Airbus Group Innovations and EOS, reproduced with permission.

[30]Press release, February 14, 2014, EOS and Airbus Group Innovations Team on Aerospace Sustainability Study for Industrial 3D Printing, http://www.eos.info/eos_airbusgroupinnovation team_aerospace_sustainability_study, (accessed May 14, 2016).

Fig. 2.15 Race car steering
knuckle produced by
DMLS[31]

Fig. 2.16 AM produced
automotive piston[32]

brake disks that are 25% lighter, with better cooling. Figure 2.16 shows an AM printed piston for automotive application. While the big attraction of AM metal processing of automotive parts remains rapid prototyping of functional test parts, the production of specialty and hard to find parts, such as those used in vintage automotive restoration is actively being pursued. Mass production of automotive parts is out of reach of current direct metal AM processes but AM methods starting with a CAD model and resulting in a metal part such as by producing a sand mold or plastic pattern is gaining wider acceptance.

Casting of large complex components can be realized by directly 3D printing a sand mold and then casting a part in metal can save development time, allowing for

---

[31]Courtesy of EOS and Rennteam Uni Stuttgart (www.rennteam-stuttgart.de), reproduced with permission.

[32]Courtesy of Beam IT, reproduced with permission.

Fig. 2.17 BinderJet
produced silica sand mold for
casting an aluminum
Formula 1 transmission
housing[34]

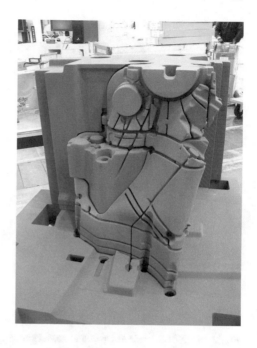

multiple design iterations during the prototyping cycle. Figure 2.17 shows a silica
sand casting mold used to cast a Formula 1 race car transmission housing using
aluminum alloy A356. ExOne provides a case study[33] where a batch size of five
castings were produced at the cost of 1500 € per part compared a lot cost of 15,000
€–20,000 € using conventional patterns, tools and lost foam casting methods. This
demonstrates that 3D printing technology can make sense for certain small lot size
casting applications.

2.7 Industrial Applications Molds and Tooling

Mold inserts can benefit from complex conformal cooling channels to speed the
molding process and improve part quality. Figure 2.18 provides a view of a model
part revealing complex cooling channels made possible by 3D printing (right view)
and the outer surface of the part produced by the DMLS process in a finished and

---

[33]ExOne case study, http://www.exone.com/Portals/0/ResourceCenter/CaseStudies/X1_CaseStudies_
All%206.pdf, (accessed August 13, 2015).

[34]Courtesy of EXONE, reproduced with permission.

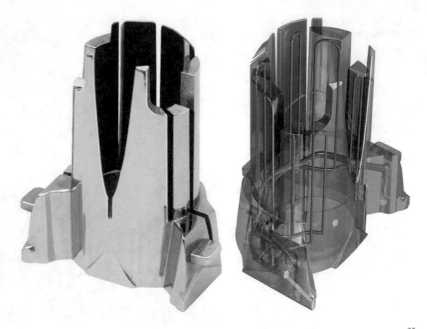

Fig. 2.18 DMLS fabricated part and model showing internal conformal cooling channels[35]

polished condition (left view). Figure 2.19 taken from a case study by GPI Prototype & Mfg. Services of the actual part, shows it was still in use after 190,000 shots resulting in a productivity increase of 48%. Applications such as these place an additional reliance on the computer-aided engineering analysis of potential designs to fully optimize the benefit of AM processing. In addition to conformal cooling, AM metal processing may be used to repair or modify existing tooling to extend the life or increase the performance of existing parts.

2.8   Remanufacture and Repair

Maintenance, repair or overhaul applications can benefit from direct energy deposition to apply coating for original parts or for repair. One such example is explained in an article "Component and Tool Life Extension Using Direct Metal

---

[35]Courtesy of GPI Prototype & Mfg Services—Lake Bluff, IL, reproduced with permission.

Fig. 2.19 Case study of the actual part in service[36]

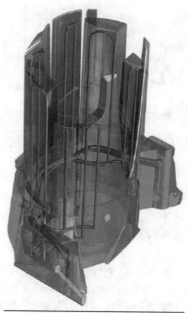

Figure 2.

Tool – Figure 2.
DMLS Tool Production: 39 Hours
Total DMLS Production Cost: $3300
Alternative Production Cost: N/A
Total Shots: 190,000 + (Still running)
Productivity Increase: 48% cycle reduction
Time to Deliver: 6 DAYS

Deposition (DMD)",[37] by Dr. Bhaskar Dutta, DM3D Technology, August 20, 2013.

Figure 2.20 shows such an example where a forging tool for a connecting rod has been coated using the DMD process, also known as directed energy deposition (DED). To overcome the heat checking and wear damages during forging, the tool was built using low-cost steel and a high-temperature Co-based alloy was applied in the heat checking areas. In contrast to the mechanical bonding of the chemical

---

[36]Courtesy of GPI Prototype & Mfg Services—Lake Bluff, IL, reproduced with permission.

[37]MTadditive.com web site "Component and Tool Life Extension Using Direct Metal Deposition (DMD), by: Dr. Bhaskar Dutta, DM3D Technology, August 20, 2013, http://www.mtadditive. com/index.cfm/trends-in-additive/component-and-tool-life-extension-using-direct-metal-deposition-dmd/, (accessed April 6, 2015).

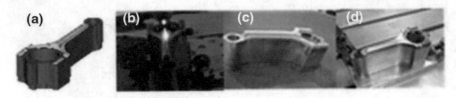

Figure 2. DMD cladding of connecting rod forging tool;
(a) CAD model showing tool base and DMD coating, (b)
DMD process in action, (c) DMD deposited tool, and (d)
finish machined tool.

Fig. 2.20 Case study by DM3D Technology of a forging tool modified with cobalt based alloy coating using the DMD process[39]

vapor deposition (CVD) and physical vapor deposition (PVD) and thermal spray coatings, DMD material is bonded to the base steel and can withstand the thermal and fatigue loading of the forging process without chipping of the coating material. DMD built hard facing material was about 6 mm thick to sustain severe forging pressure and also allow for machining of the tool multiple times. DMD applied tools had four times longer life over conventional tooling and resulted in significant cost savings while reducing downtime. In another example[38] Optomec demonstrates repairing an impeller blade using laser beam directed energy deposition (Fig. 2.21).

2.9 Scanning and Reverse Engineering

Scanning technology can use laser or photographs to capture an object's shape and use reverse engineering software to recreate a model of the object. That model can be used to 3D print plastic patterns or sand molds of direct to metal parts.

---

[38]Impeller repair article, the fabricator, http://www.thefabricator.com/article/metalsmaterials/fabricating-the-future-layer-by-layer, (accessed August 13, 2015).

[39]Courtesy of MTAdditive and DM3D Technology, reproduced with permission.

Fig. 2.21 Directed Energy Deposition Repair of Impeller Pump[40]

Figure 1
In this directed energy deposition process, a laser deposits layers of metal to repair an impeller pump. Photo courtesy of Optomec, Albuquerque, N.M.

Fig. 2.22 Example of scanned parts to CAD model[43]

Geomagic,[41] owned by 3D Systems, provides hardware and software solutions to allow 3D scanning and the creation of 3D models to be used in original and reverse engineering applications.[42] One example, in Fig. 2.22, demonstrates the ability to

[40]Photo courtesy of Optomec (reproduced with permission); LENS is a trademark of Sandia National Labs.

[41]Geomagic Web site, http://www.geomagic.com/en/, (accessed March 26, 2015).

[42]Geomagic case study, http://www.geomagic.com/en/community/case-studies/rebuilding-a-classic-car-with-3d-scanning-and-reverse-engineerin/, (accessed March 26, 2015).

[43]Courtesy of 3D Systems, reproduced with permission.

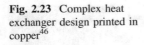

Fig. 2.23 Complex heat exchanger design printed in copper[46]

scan motorcycle engine parts, process the point cloud data into a model that can be features and assembled into a 3D model that can also be used to 3D print a plastic or metal component. The software can interface with professional level computer-aided design (CAD) software such as Catia, Solidworks, etc.

2.10 Software

Software used to create complex designs is being created by cutting-edge companies such as the company WithinLab (now Autodesk Within). The software is used to assist in the design process for complex internal structures, repeating structures, variable density structures, and complex shapes, as needed, helping to realize the potential of AM designs, such as the complex heat exchanger design built in copper as shown in Fig. 2.23. Siavash Mahdavi of Within has a TED talk[44] and the Web site[45] provide videos that explain the technology that can create and optimize latticed microstructures and surface structures. The software adds an extra layer of file encryption enhancing the security of intellectual property.

---

[44]Siavash Mahdavi TED talk, posted March 13, 2012, http://tedxtalks.ted.com/video/TEDx Salzburg-Siavash-Mahdavi-St;Featured-Talks, (accessed April 6, 2015).

[45]Within Web site with videos and software for medical implant design, http://withinlab.com/overview/, (accessed April 6, 2015).

[46]Courtesy of 3T RPD, reproduced with permission.

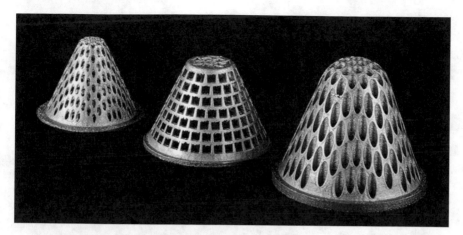

Fig. 2.24 Complex filter designs produced by AM[49]

2.11 Engineered Structures

Complex internal features, such as turbulators to increase heat conduction in cooling channels, combined with conformal cooling channels in molds and mold inserts, allow designers unprecedented freedom in optimizing the thermal or mechanical functions of a part while offering shorter cycle times. Air ducts with laminar flow design, complex heat exchangers with repeated sub-elements and other complex structures, are being demonstrated. The benefits of complex internal structures may outweigh the need for accuracy or surface finish of these structures. One interesting application is associated with filtration technology[47] as shown in Fig. 2.24. AM metal technology was used to create complex filter components offering high flow rates and greater efficiency reducing operating costs. In another example, engineered surfaces at the microscale can improve coating adhesion of interfaces between AM produced titanium hardware and carbon fiber structural members. Figure 2.25 shows a cut away view of the complex internal structure of a gas emissions rake displaying what is made possible by using AM design and fabrication. New applications and complex shapes are being reported every day in a growing number of new industry publications, such as Metal Additive Manufacturing,[48] demonstrating the expanding capability of the technology and those who use it.

---

[47]Additive Manufacturing-What you need to know, Filtration + Separation.com article, February 20, 2014 article, http://www.filtsep.com/view/37036/additive-manufacturing-what-you-need-to-know/, (accessed March 28, 2015).

[48]Metal Additive Manufacturing, Spring, 2015, Vol. 1, No. 1.

[49]Courtesy of Croft Additive Manufacturing, reproduced with permission.

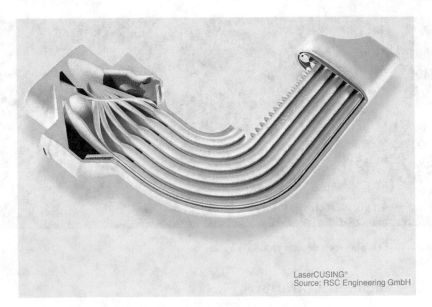

LaserCUSING®
Source: RSC Engineering GmbH

Fig. 2.25 Cut-away view of a gas emissions rake with complex internal structure fabricated using LaserCUSING®[50]

2.12 Functionally Graded Structures and Intermetallic Materials

Ongoing research at universities and government research laboratories demonstrate the ability of AM metal processing to locally affect and potentially control the microstructure of AM deposited metal using the wide range processing parameter available to AM. Electron beam additive manufacturing has been demonstrated to enable site-specific control over the microstructure within an AM metal deposit[51] offering the potential for engineering the metal properties of localized regions and features of a part. The metallurgy and processing science of metal additive manufacturing is an active area of study.[52] An in-depth understanding of the AM metal

---

[50]Courtesy of Concept Laser GmbH, RSC Engineering GmbH, reproduced with permission.

[51]R.R. Dehoff, M.M. Kirka, W.J. Sames, H. Bilheux, A.S. Tremsin, L.E. Lowe and S.S. Babu: 'Site specific control of crystallographic grain orientation through electron beam additive manufacturing', Mater. Sci. Technol., 2015, 31, (8), 931–938.

[52]W.J. Sames, F.A. List, S. Pannala, R.R. Dehoff & S.S. Babu (2016): The metallurgy and processing science of metal additive manufacturing, International Materials Reviews, http://dx.doi.org/10.1080/09506608.2015.1116649, (accessed May 14, 2016).

Fig. 2.26 a Custom sternum chest implant.[53] **b** Sternum chest implant illustration.[54]

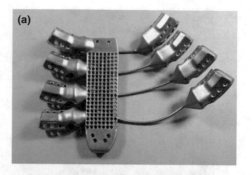

(a)

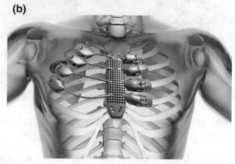

(b)

processes, the chemistry and metallurgy resulting from these processes, combined with a first principal understanding of the metallurgy and physics may offer simulation tools for the prediction and design and for controlling and engineering the resulting properties and performance of AM produced parts. While localized control of the micro structure within an AM produced part is still at the research stage, localized control of the structure of the deposit at very small scales is made possible by AM and can be used to functionally grade the materials such as by adding a coating or additional layers of a different material. Grading or changing the structure of the deposit, such as within medical implants, can locally change the part function from that of a load bearing member to a region that fosters bone ingrowth. Figure 2.26a and b shows a titanium prosthetic sternum or rib cage fabricated using the PBF-EB process. It was implanted into the chest of a 54-year old cancer patient who lost his sternum and four ribs with the removal of a large tumor. Engineers at Anatomics in Melbourne, Australia used the patient's own CT scans to engineer the custom shaped device. The perforated sternum portion provides rigid strength while the four thin rods are designed to flex during breathing.

---

[53]Designed by Anatomics Pty Ltd., Melbourne Australia, reproduced with permission.

[54]Designed by Anatomics Pty Ltd., Melbourne Australia, reproduced with permission

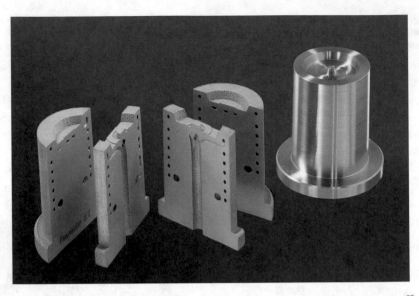

Fig. 2.27 Tool insert with internal cooling structures made from Hovadur K220 by SLM.[57]

A cutting-edge example of using a DED laser based system has been demonstrated by Peter Dillon at the Jet Propulsion Laboratory (JPL), using materials such as A286, 304L, Invar36 clad, in a collaboration with Penn State University (PSU), NASA, JPL, and Cal Tech Pasadena. GE Aviation's Avio Aero is working to use TiAl, an intermetallic material offering unique properties and performance, in turbine blades and JPL for functionally graded piping using the EBM electron beam melting process.[55]

2.13   Technology Demonstration

A research team at Fraunhofer ILT in Aachen has demonstrated the use of SLM for the deposition of copper tooling inserts with internal cooling channels,[56] Fig. 2.27. In the InnoSurface project, funded by the German Federal Ministry of Economics and Technology, the team has succeeded in modifying the SLM process by

---

[55]Metal Additive Manufacturing article, August 20, 2014, http://www.metal-am.com/news/002896.html, (accessed March 26, 2015).

[56]Prototype Today article, http://www.prototypetoday.com/fraunhofer/components-made-from-copper-powder-open-up-new-opportunities, (accessed March 28, 2015).

[57]©Fraunhofer ILT, Aachen/Germany, reproduced with permission

Fig. 2.28 The Solid Concepts 3D printed 1911 pistol[59]

increasing the power from 200 to 1000 W and tailoring the laser beam profile, changing the inert gas control system and mechanical equipment to accommodate the high reflectivity and thermal conductivity of copper to improve laser coupling and melting. A deposit density of near 100% was reported.

In the technology demonstration[58] shown in Fig. 2.28, a working 1911 design type firearm was 3D printed in solid metal and test fired to prove the concept. The company, Solid Concepts had produced other types of the handgun as a special issue then sold to consumers.

2.14 Hybrid Additive/Subtractive Systems

The integration of AM processing with advanced subtractive processing such as milling or turning is another area of technology development promising to leverage the best of both worlds. A new world of design opportunities exist in the marriage of two or more processes on the same build platform allowing a hybrid design relying on the strengths of each process.

Commercial systems integrating laser directed energy deposition within a precision machining platform have reached the market place. They offer the capability to add complex features or surfaces to simple base shapes or add to complex shapes

---

[58]Solid Concepts blog, https://blog.solidconcepts.com/industry-highlights/1911-3d-printed-guns-will-sell-lucky-100/, (accessed August 13, 2015).

[59]Courtesy of Solid Concepts Inc. under CC BY-SA 4.0: https://creativecommons.org/licenses/by-sa/4.0/deed.en.

already formed by CNC machining. DED-L is well suited for this application as laser powder feed heads can be made small enough to fit within the confines of these systems. These hybrid systems are being marketed as solutions for small parts needing complex features made from hard to machine materials or large workpieces with high stock removal volumes. One such hybrid machine the LASERTEC AM/SM system[60] made by DMG Mori, allows AM feature deposition and conventional milling within the same setup.

Another competing approach to hybrid systems incorporates a multi-turret milling platform into a PBF-L system to mill surfaces and contours deposited by PBF-L in an attempt to attain the desired accuracy while the part is being built. Other systems have demonstrated the use of robotically controlled tools to combine laser cladding with five-axis machining, in process measurement, polishing, annealing and cleaning all into one system setup.

The Hybrid Manufacturing technologies[61] Web site has a video that shows machining, laser cladding, on-machine gauging and post-machining operational sequencing all on one machine for the repair of a turbine blade as part of the UK RECLAIM project.

Lumex Advance-25 hybrid machine[62] by MC Machinery Systems, in partnership with Matsuura Machinery Corporation has created the LUMEX Avance-25 metal laser sintering hybrid milling machine combining metal laser sintering (3D SLS) technology with high speed milling technology, enabling one-machine, one-process manufacturing of complex molds and parts.

2.15 Key Take Away Points

- 3D printing and additive manufacturing of metal prototypes is now widely adopted with new applications being show cased across a wide range of industrial sectors.
- Custom, on demand, one of a kind personal items such as jewelry or consumer goods are now being offered commercially.
- Medical devices, surgical aids, and implants are being approved for use and can be matched to a person's anatomy. Unique surfaces and lattice structure offer benefits for bone ingrowth and biological integration.

---

[60]DMG Mori LASERTEC web link, http://us.dmgmori.com/products/lasertec/lasertec-additivemanufacturing, (accessed December 18, 2016).

[61]Hybrid Manufacturing Technologies Web site and video, http://www.hybridmanutech.com/, (accessed April 8, 2015).

[62]MC Machinery Systems, http://www.mcmachinery.com/whats-new/Matsuura-Lumex-Avance-25/, (accessed April 8, 2015).

- Dental devices with small lot sizes, high precision, and high value products are seeing wide adoption.
- AM Engineered aerospace components with complex internal structures, cooling channels, intricate lattice, and honeycomb features have shown the potential for significant energy savings, and the benefit of lightweight strong structures. Multiple conventionally produced components may be combined into a single AM part significantly reducing part count while cost savings are realized by improvements to the buy-to-fly ratio for expensive advanced materials.
- Industrial tooling and molds offer improvements to conventional production lines while repair and remanufacturing applications are improving and extending the life of legacy systems.
- Production of high-cost, low-volume components is focused on materials that are costly or difficult to process by conventional methods.
- Hybrid machines are leading the way to integrate AM capable machines into the digital factory.

Chapter 3
On the Road to AM

Abstract The advantages offered by additive manufacturing intersect a wide range of industries, occupations, and potential users. This chapter provides various scenarios of the artist, student, inventor, business owner, engineer, or technology manager describing how additive manufacturing can combine with their skills, capabilities, and interests to assist the reader to place themselves within the context of this new technology. This chapter provides an introductory overview of the types and capabilities of current additive manufacturing systems that can best meet your application. A discussion of market and technology drivers asks key questions to assist the reader in clearly defining their needs and how the advantages or limitations of AM technology can drive the decision of whether or not to adopt AM technology. Questions are presented to address considerations such as material cost savings, energy and time efficiencies, and conversely what legacy drivers can impede the reader's adoption of AM. Additive manufacturing of metal exists at the confluence of multiple diverse technologies with a language of terms borrowed from each. The reader is introduced to these terms and with the high level overview provided by this chapter, prepared to build an understanding of those technologies and materials fundamental to AM metal processing.

3.1 You are Here →

If you are an artist, you may be in the jewelry business or creating free form sculpture either directly or from using computer models. Your inspiration may be derived from nature, perception, emotion, or be inspired by others. You may create single works, small lots series, or cater to the crafts market. You may have heard 3D printing can serve specialty markets with global Web-based reach, either through direct sales or commissioned works. 3D print on demand offers the potential to reduce inventories of your works and in the case of precious metals, minimize the usage, storage, and inventory. Computer model-based artwork will allow you scale and modify existing designs and with the availability of low-cost 3D modeling software and a desk computer. The wide range of professional 3D print service

© Springer International Publishing AG 2017
J.O. Milewski, *Additive Manufacturing of Metals*, Springer Series
in Materials Science 258, DOI 10.1007/978-3-319-58205-4_3

providers offer you a wide range of options allowing you progress from plastic, wax, and sand molds to direct metal including platinum and gold.

If you are student with an interest in technology, you are well aware of the exciting developments on the cutting edge of science and engineering with exciting new computer applications, games, robots, drones, Internet of Things, artificial Intelligence, and 3D printing. Your inspiration is derived from hacking and mashing together the latest technology, but also to take fun and interesting courses. You are free to dream far outside the box, into virtual worlds, and into the future. More and more schools, educational institutions, colleges, and universities are gearing up their facilities and hiring instructors to help create your vision for the future, not to mention future employment.

If you are an inventor or a do-it-yourself type, you feel empowered by the possibility offered by 3D printing and additive manufacturing metal. The reduced reliance on manufacturing infrastructure allows persons without access a wide range of conventional metal sources and metal processing facilities, to design and have fabricated complex metal parts, in some cases those that cannot be made any other way, and have them delivered to your door faster than ever dreamed possible. The availability of low-cost software for modeling, scanning, and access to 3D printing services offers the possibility for you to design your very own personalized objects unique to one person.

If you are a business owner fabricating or manufacturing metal components, or serving the metals market you have heard about additive manufacturing and rapid prototyping using metals. How does it work? What metals can be 3D printed? What are the best applications and which way are the markets going? What is your competition doing and what advantages could additive manufacturing of metal offer my business? What does it take to get started and where do I learn about it? What resources are available to me and how much will it cost to find out?

If you are an engineer in a technology corporation that is adopting or exploring additive manufacturing of metal components, what are the career path opportunities? What is your skill set and how do those fold into the multidisciplinary field of additive manufacturing. How can your expertise in computers, software, design, metallurgy, or metal fabrication fit into the big picture? How does a high-powered team of technically diverse experts pull together, energize, and create a new future? What are the technical challenges faced by AM and how does one manage this rapidly changing world of technology?

Given the above examples, what is in your backpack? What are your existing capabilities, capacities, and skills related to AM and your specific needs? Given your AM needs, how much of these capabilities exist in-house? What would be the cost to establish them? What would be the cost of outsourcing? Consider the following points when assessing your current capabilities with respect to the adoption of AM.

- Modeling, design, software, and engineering capability and skills
- 3D printing or rapid prototyping experience with nonmetals
- Conventional metal fabrication skills and capability

- Relevant materials or metallurgy experience
- Established use of relevant AM and CAD design resources
- In-depth understanding of a relevant AM target market
- Floor space and facility configuration
- Capital investment resources.

Next we introduce and summarize the range of AM processes as the vehicles that can take you to your first AM destination and guide you to greater detail provided later in the book.

3.2 AM Metal Machines, the Vehicles to Take You There

This section provides a quick introduction and high-level overview of the type of AM metal processes, how they work, what they are used for and what advantage they provide (Fig. 3.1). We provide this high-level overview to assist the reader to zoom in on process detail provided later in the book and focus on the processes that best fit your application.

Laser Beam Powder Bed Fusion, (PBF-L) is the most widely applied and perhaps the most evolved AM metal technology. A range of metal alloys are available but generally limited to those engineering metal alloys optimized for powder bed fusion AM. The high cost of powder is currently a limiting factor to its adoption. The current geometric surface model representation, described later as the STL model, has simplified and speed the adoption PBF-L. Based on 3D printing technology, the surface model is sliced into planar layers which are used to define the path of a laser beam that is scanned to fuse the layer shape into the powder bed. An additional layer is applied and fused, layer by layer forming the part. Additional

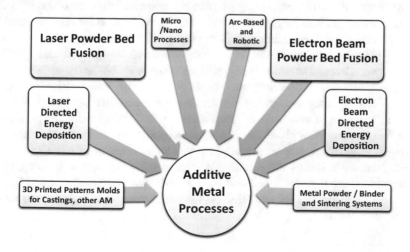

Fig. 3.1 Additive manufacturing metal processes

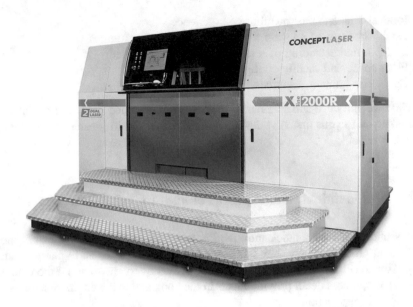

Fig. 3.2 Concept Laser X2000[1]

support structure must be added to the design to enable building overhangs and downward facing surfaces, increasing the skills needed to complete the process design. Current build volume sizes are limited, on the order of a 400 mm × 400 mm × 800 mm, with deposition rates on the order of 5–20 cm^3/h, but process developments are continually improving these capacities and build rates. As deposited surface quality is the best of all current AM metal systems and is related to powder particle size typically ranging from 10 to 60 μm. High-purity argon inert environments allow the processing of reactive materials such as titanium, while nitrogen generators may be used to provide a lower cost option for certain materials. Some vendors offer open architecture, allowing greater access to process parameters and machine interfaces, assisting in the development of certified and qualified process procedures. Distortion and residual stress can be an issue with some materials while post-processing, such as heat treatments and hot isostatic press (HIP) processing, may be required to fully develop the desired properties. Systems range in price from hundreds of thousands of dollars, to millions of dollars, depending on size, laser power, and optional capabilities such as powder recycling or system diagnostics. Expect to spend up to a million dollar or more to set up a PBF-L fabrication facility not counting post-processing equipment. Concept Laser has developed the LaserCUSING system, a PBF-L type process shown in Fig. 3.2. The leading vendors of PBF and DED AM systems along with their process names are shown in Table 3.1.

---

[1]Courtesy of concept lase GmbH, reproduced with permission.

Table 3.1 The leading vendors of PBF and DED AM metal systems and process names

Manufacturer	Process	Process name	ASTM designation
EOS	DMLS	Direct Metal Laser Sintering	PBF laser
Concept Laser	LaserCUSING®	LaserCUSING	PBF laser
SLM Solutions	SLM	Selective Laser Melting	PBF laser
Renishaw			PBF laser
Realizer			PBF laser
Arcam AB	EBM®	Electron Beam Melting	PBF Electron beam
DM3D	DMD®	Direct Metal Deposition	DED laser
Optomec	LENS®	Laser Engineered Net Shape	DED laser
RPM Innovations	LDT	Laser Deposition Technology	DED laser
Sciaky	EBAM™	Electron Beam Additive Manufacturing	DED electron beam

Electron Beam Powder Bed Fusion (PBF-EB) is a powder bed process that uses an electron beam heat source to fuse the powder. A range of metal alloys are available but limited to those powders optimized for AM PBF-EB use. Advantages of the process include a higher temperature deposition environment with powders typically heated to ~700 °C, with the benefits of reducing residual stress and distortion. The process is performed in a vacuum environment offering benefits to material such as titanium and cobalt chrome alloys used in critical applications. The materials must be conductive imposing limitations on material selection. As with PBF-L, the dependency on the STL file format exist, but design and use of support structures is reduced as the surrounding powder is lightly sintered to help form the support. This pre-sinter feature, reduces the need for extensive support structures, may also simplify the loading and stacking of multiple parts within a single build volume and build cycle, maximizing system productivity. Build volume dimensional sizes of approximately $350 \times 350 \times 380$ mm and build rates up to 80 cc/h, placing limits on the building of larger parts. The process relies on the use of larger powder size, on the order of 45–105 μm, ultimately resulting in a somewhat rougher surface finish when compared to PBF-L. The electron optics offer a more rapid control of the electron beam raster and focal conditions compared to PBF-L, new EB gun development has expanded the beam control capability and beam power. As with laser powder bed systems, they can range in price from hundreds of thousands of dollars to over one million dollars depending on the size, EB power, and optional capabilities such as powder recovery or system diagnostics. Arcam AB has developed and is currently the sole provider of the Electron Beam Melting (EBM) process. An EBM machine is shown in Fig. 3.3.

Laser Beam Directed Energy Deposition, (DED-L) has evolved steadily over the past 20 years as an outgrowth of laser cladding and CNC motion technology. A number of laser/powder deposition heads are available and are being interfaced with a wide variety of lasers and purpose built systems. Cladding heads for internal pipe bores or modular systems for upgrading or repurposing of existing CNC machine systems are available. CNC machine tool and DED-L hybrid machines have been demonstrated and are now being offered commercially. Advantages over

Fig. 3.3 Electron beam melting machine[2]

PBF systems include the cladding and coating of complex 3D surfaces for repair and location-specific coatings. A wider build envelope on the order of 1500 × 900 × 900 mm allows for the processing of larger parts, while deposition rates can reach 500 cm³/h. A wider range of metal powder alloys is also available, as the powder requirement for particle size and shape is less stringent than for PBF systems, resulting in lower costs. Multiple powders can be fed or sequenced allowing the deposition of functionally graded deposits. Atmospheric or high-purity inert chambers are featured. Layer-by-layer deposition (also known as 2½ D) is based on STL file input with the same benefits and limitations as PBF-L. Full CAD/CAM/CNC software support, offering a fully parametric path from design to tool path is possible. Vendors offering open architecture, allow greater access to process parameters and machine interfaces, assisting in the development of certified

---

[2]© Arcam, reproduced with permission

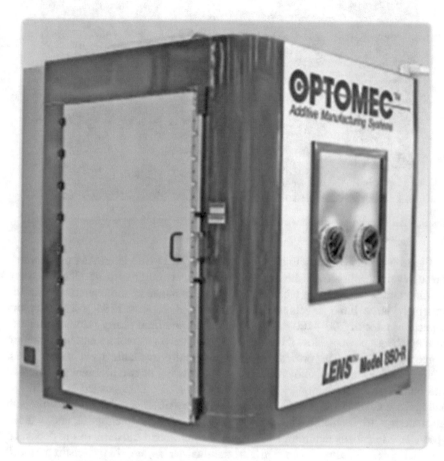

Fig. 3.4 LENS 850 R[3]

and qualified process procedures. Distortion and residual stress can be an issue with some materials while post-processing to include heat treatments and HIP processing may be required to fully develop the desired properties. Systems range in price from hundreds of thousands of dollars to repurpose or retrofit a CNC system to over $1 M for a fully featured DED lab depending on the size, laser power, and optional capabilities such as powder recyclers or system diagnostics. The Optomec LENS 850R DED-L machine is shown in Fig. 3.4.

---

[3]Photo courtesy of optomec (reproduced with permission); LENS is a trademark of Sandia National Labs.

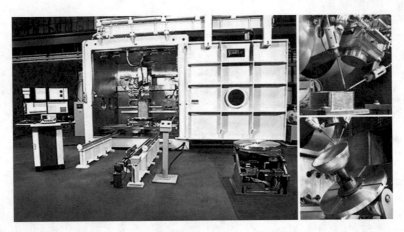

Fig. 3.5 An electron beam additive manufacturing (EBAM™) 110 system from Sciaky, Inc.[4]

Electron Beam Directed Energy Deposition, (DED-EB) is an AM process that has evolved steadily over the past 20 years as an outgrowth of electron beam welding and weld cladding technology. Big advantages of the technology are the extremely large build envelopes on the order of $1854 \times 1194 \times 826$ mm and deposition rates of 700–4100 cm³/h. In addition, the high purity vacuum chamber and high beam powers allow the processing of reactive, refractory and high melting temperature alloys. The process uses commercially available weld wire shapes, offering a wider range of alloys and allows the feeding of two different materials in one build cycle. Deposition is typically limited to 2½ D deposited shapes originating from CAD/CAM and CNC control software offering a fully parametric path from design to toolpath. The process features a large melt pool and welded bead-shaped deposits requiring additional machining and finishing of the deposited near net shape buildup. Distortion and residual stress are characteristic and will require post-processing including stress relief or other heat treatments to fully achieve the desired properties. Expect to spend well over one million dollars to set up a fully featured DED-EB facility not counting post-processing capability. A Sciaky EBAM machine is shown in Fig. 3.5. EBAM features a closed-loop control system IRSS® (Interlayer Real Time Imaging and Sensing System).

Gas Metal Arc Weld and Plasma Arc Directed Energy Deposition, (DED-GMA, DED-PA) and other arc-based systems are typically used for cladding and repair, or shaped weld deposit buildup. They can utilize a wide range of weld wire alloys. The extensive heat buildup, characteristic of the process, often requires a large inert gas shielding chamber to process reactive or refractory alloys. Recent developments in control systems and robotics have to fully integrate this process with CAD/CAM control to produce large, complex, near net shaped components, such as those achievable using DED-EB with similar high deposition rates and large build

---

[4]Photo courtesy of Sciaky, Inc., reproduced with permission.

Fig. 3.6 ExOne binder jet machine[5]

envelopes. Robots allow complex deposition with full 3D motion paths. As with DED-EB, a large melt pool and weld bead-shaped deposits typically require 100% machining and finishing of the deposited near net shape buildup. Extensive distortion and residual stress is characteristic and may require post-processing, including stress relief and heat treatments to achieve the desired properties. Arc-based system may range from tens of thousands of dollars to hundreds of thousands of dollars or more depending on the energy source and level of automation.

Binder Jet Technology offers many of the benefits and properties of metals by creating composite structures, often exceeding the properties of other 3D printed composites when using plastic or polymer matrix materials. Examples include the bonding of stainless steel, iron, and tungsten using a lower melting point matrix metal such as bronze. While the selection of materials is limited, process development as applied to other metals is ongoing. One of the greatest benefits include the direct binding of sand cast molds and has been demonstrated for very large metal castings. Accuracies can be on the order of powder bed fusion methods but with much larger build envelopes. Advantages include the processing of materials not easily melt processed. Properties of bronze infiltrated sintered metal will not match those of wrought alloys but in certain cases hot isostatic pressing can densify a wide range of engineering alloys without the infiltration step. An ExOne binder jet technology machine is shown in Fig. 3.6.

CAD to Cast Metal using wax or polymers offers an alternative way to use 3D printing technology to realize fully metal shapes. These processes use 3D printing of plastics, polymers, or wax to produce a pattern from a 3D CAD model. The

---

[5]Courtesy of ExOne, reproduced with permission.

pattern may then be used to create sand or investment molds for metal casting. Advantages include a lower cost alternative to PBF or DED metal systems but disadvantages include limits in material selection and part design due to the casting process. It is attractive for both small components, such as jewelry, and components larger than typical PBF build volumes. The advantages of complex design and structures attainable by PBF may not be realized. A simple apparatus for casting aluminum combined with a low-cost wax printer may configure a capability for under $5K.

Ultrasonic Consolidation, and Sheet Lamination of metals is a solid state joining process that typically uses an ultrasonic tool to bond sheet or strip layers of metal or other composite materials into a consolidated shape. To realize the final part, milling or machining operations are required to remove the part from the base plate and remove the bulk of unfused layers surrounding the solid region. Aluminum and titanium have been joined but not all metals may be used with this process. These processes can solid state bond dissimilar metals in effect functionally grading the performance of the part. Geometry and part sizes are limited but materials that cannot be melt processes may benefit from this process. Cold Spray Forming is another solid state joining process where particles are delivered using high-velocity deposition to spray form and buildup material layer by layer.

Micro and Nano scale methods and applications are highly specialized and mostly in research and development. They may see wider use and adoption due to publicity received by the recent industrial interest in AM. Technology to ink jet nanoscale-coated metal particles is being developed to form metallic components.

AM service providers exist to assist users, across the entire spectrum of AM metal processing, from 3D design to fabrication. These services offer an attractive way to evaluate candidate AM technology prior to making the commitment and investment to establish an in-house AM capability. Training and consulting services specific to AM technology disciplines such as metallurgical engineering, or computer programming, are already being established to augment the talent pool of an established AM capability.

3.3 Market and Technology Drivers

Here we introduce AM technology drivers. What motivates users to start the journey toward AM technology adoption? What makes AM so attractive? What motivates users not to start the journey or take a good hard look and wait? What markets could benefit the most from AM? What potential energy or cost savings may be realized? What efficiencies or optimization may result in savings all along the value chain? What investment is needed? What legacy driver can impede or slow the adoption of AM? What design freedom can take you there?

Throughout this book these drivers will be tied to materials and processes as we delve deeper into the technology and each process. We will elaborate on these drivers and extend these topics, as the answer to these questions will depend on the application, the material and the specific AM process. Some of these answers will drive one's decision, positive or negative, regarding AM adoption.

Top AM Market Drivers:

1. The markets most attractive to AM in terms of products, materials, processes, and applications are described in Chap. 2. What others exist?
2. What conventional fabrication processes are used to serve that market today? Which conventional processes are easy and which are difficult, costly or time-consuming?
3. Who are the market's customers? What are their primary needs? How mature is the market?
4. What markets could benefit most from customization or individualization?
5. Which markets rely on small lot manufacturing of high-value components?
6. Which markets rely on mass production that uses specialized tooling that may be optimized by fabrication using AM?
7. What market utilizing specialty metal components are rapidly changing with potential for expansion?
8. What is the potential for individualization within the market?
9. Where does investment within the value stream need to be made? Where will return on investment be realized and when?
10. What would be the impact of Additive manufacturing (AM) adoption farther upstream or downstream within the existing value chain?

Top Material, Cost, and Efficiency Drivers

1. What material properties and performance data exists for AM metals? Is this data adequate for engineering designs or the manufacture of certified components?
2. What processes are currently used for difficult to fabricate metals and what are their limitations? Can AM overcome these conventional limits?
3. What materials are difficult to recycle in the conventional processing stream? Can AM reduce scrap and waste streams of conventional processing?
4. Will AM part costs in $ per kg provide benefit? What is the balance of material savings versus increased AM material costs?
5. What are the limits to AM material types, both powder and wire forms, in terms of cost and availability?
6. What are the potential energy savings from a complex AM design and part? Can there be fuel or energy savings, thermal management, or improved flow characteristics?
7. How much value exists within an optimized design and prototype cycle?
8. Can the mechanical performance of a design be optimized using AM materials and processes, such as strength, hardness, or wear characteristics? Can AM-related designs result in long life components, particularly those serving in harsh environments?
9. What is the cost of downtime when compared to the cost of a rapid replacement, on-demand AM produced part? How difficult are spares to procure and

how long does it take? How remote is the potential demand from the supply of replacement parts, such as on a sea going vessel?

10. What is the opportunity for and how much value may be realized by combining multiple parts or part functions into one component?

Top Legacy Drivers

1. How effectively will AM parts interface with legacy system design constraints?
2. What are the failure modes of existing components? Can AM be used to produce a better performing part using repair or remanufacture?
3. Can AM methods combined with reverse engineering replace legacy or obsolete parts for an aging infrastructure?
4. How cost effective is performing reverse engineering or re-engineering of legacy components?
5. What legacy knowledge for existing components or processes exists? Can this be easily converted to model-based definition and stored in the cloud for retrieval allowing AM regeneration of parts on demand?
6. Will loss of legacy knowledge, existing product definitions, or limitations to design definitions impede archival access?
7. What legacy parts or part producers are no longer available?
8. Can AM serve the market tail of products entering the end of their life cycle?
9. How do you manage the mismatch of legacy workforce skills while upgrading to AM skills?
10. How do you balance legacy thinking while managing outside the box thinking to achieve AM results?

Top Design Drivers

1. What can be the benefits or drawbacks of increased AM design freedom or complexity?
2. What software upgrades, skills, and costs are required to achieve these complex designs? How mature are these software platforms?
3. What improvement may be realized by combining, adding, or improving the function of each part feature?
4. What additional AM-specific features are required to assist in manufacturing, finishing or inspecting an AM part?
5. If less volume of a higher performance material could be used, would you design for it?
6. What surface finish is achievable as AM fabricated or as a result of post-processing?
7. What part distortion or internal stress can result from AM processing? How are these controlled?
8. What are any downstream production scaling considerations?
9. What is the status of AM parts complying with certification standards or regulations?

10. How can secondary processing requirements such as CNC machining, heat treatment, and surface finishing affect AM design decisions?

Top AM Process Drivers:

1. What are the inherent limitations of the various AM processes?
2. What are the various AM system size, speed, and capacity limitations?
3. What is the best internal or external part finish that can be expected?
4. What design factors can alter this surface condition?
5. What secondary and post-process operations are required?
6. What inspection methods are best for AM produced parts? What are the limitations?
7. What are the deposition rates for each process? Which are faster and which are more accurate?
8. What is the accuracy and repeatability of the various AM processes, from part to part and machine to machine?
9. What are the facility requirements and safety envelope for an AM processing shop?
10. What are the considerations for operational costs and maintenance?

After reading this book and following the links provided to the latest developments, your newly acquired AM knowledge will enable you to survey the most current AM system hardware, software solutions, post-processing capability, applications, and required skills to determine how well you are positioned to best use AM processing. Structuring your decision process can assist by mapping your markets and designs with available materials and existing AM platforms, while factoring in your existing capabilities and skills, and ranking the options.

3.4   A Pocket Translator: The Language of AM

One of the barriers to entry into the AM world is the multidisciplinary nature of the technology. Designers, computer modelers, laser welders, motion control technologists and metallurgists, to name a few, all converge in the broad field of AM metal. They all bring their technical terms, definitions, acronyms, and technical slang to the table and do their best to communicate. This alphabet soup (Fig. 3.7) can be daunting for the new comers and given the rapid growth of the technology there are many new comers. In light of this, particular attention is paid in this book to calling out new terms in italics, identify acronyms at first use, and providing a detailed index to reference. The reader is encouraged to review the Acronyms presented at the beginning of the book and locate the Glossary and Index sections at the back of the book as you will reference these sections along the way.

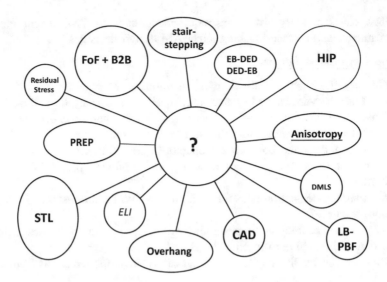

Fig. 3.7 AM metal borrows terminology from a wide range of technical fields

3.5   Key Take Away Points

- The best way to determine how additive manufacturing can best work for you is to begin with a high-level view of your needs, skills, and interests and match them with the AM process and materials and which is right for you.
- A survey of what AM technology and machines are currently available for producing metal parts will help you put into perspective your needs and what new AM opportunities are now within your reach.
- A review of the market and technology drivers for AM adoption will help the reader ask the most relevant questions clarifying the target technologies offered by AM, what they offer, where the potential user currently resides and to choose the best path forward.
- AM metal processing borrows technical terms from a wide range of technologies ranging from computer modeling, to metallurgy, to machine control. This book provides and extensive listing of terms, jargon, and acronyms needed by the user as a reference and to technically communicate and further investigate this rapidly evolving technical field.

Chapter 4
Understanding Metal for Additive Manufacturing

Abstract Additive manufacturing of metal exists at the convergence of a wide range of advanced technologies ranging from the design of computer solid models and computer-driven machines to high-energy beam processing of materials. Many technologists with diverse backgrounds are being drawn into additive manufacturing with little knowledge of or experience working with metals. This chapter provides a quick refresher of the building blocks of metal, its crystal or microstructure and the properties resulting from the chemistry of the specific metal alloy and the process used to form it into useful shapes. Simple examples are used to illustrate the wider range of forms and structures metal can take as a result of operations such as casting and rolling, then compared with metal processed by additive manufacturing methods. The non-metallurgist is introduced to less common forms of metal such as metal powders, wire, and electrodes as well as the microstructures formed by sintering and solidification. Less common materials such as composites, intermetallic, and metallic glasses are introduced as the potential use and application of these advanced materials is increasing as made possible by the adoption of additive manufacturing.

Metal: what is it? What properties does it have that are relevant to building 3D printed objects? What metals work best for additive manufacturing and why? Where do you start to select which metal is best for your application. This chapter is an important stepping stone to your understanding and success in additive manufacturing metals.

Many people without any engineering background know very little about metal aside from common uses and names such as steel, aluminum, copper, chrome, and gold. To apply the additive manufacturing metal printing processes successfully, the user needs a base level understanding of how metal and especially *metal alloys* result in the properties and performance we take for granted in everyday objects. This chapter offers an introduction to the language of metal, using simple descriptive terms in the context of additive manufacturing metals allowing the reader to understand what metals are best suited for AM use and why. We discuss how AM may affect metal properties and part performance when compared to using

© Springer International Publishing AG 2017
J.O. Milewski, *Additive Manufacturing of Metals*, Springer Series in Materials Science 258, DOI 10.1007/978-3-319-58205-4_4

conventional metal processing methods. References and the Web links are provided as learning pathways to be followed if more information is required for a specific term or material. This chapter sets a foundation for understanding technical terms and descriptions of metal within the book, vendor literature, technical articles and content across the web. One example of Web-based learning is provided by the Materials Science and Technology (MAST) Teacher's Workshop prepared by the Department of Materials Science and Engineering, University of Illinois Urbana-Champaign.[1]

To illustrate, we will start with some of the properties common to all metals and give a few examples as they relate to AM. A basic understanding of these will help you understand how metal responds to being melted, fused or sintered. We will also touch upon metal powder and some less common alloys. Later in the book we will provide additional detail on how metal responds to each of the main AM metal processes and why you should care. But for now, let us discuss the basics.

4.1 Structure

4.1.1 Solid, Liquid, Gas, and Sometimes Plasma

Most folks think of metal as a solid piece of stuff that is hard, strong, heavy, and durable. Often they associate a color with it and are aware it can be melted, as they have all seen the movie where the death ray has reduced the Sherman Tank to a glowing blob or when a strange shape shifting alien (think Terminator movie) turns into a shiny liquid metal form and then back into a human-like form. Technical folks and metallurgists know there is more to it than that.

We all know matter has three states or phases; *solid, liquid, and gas*. As an example, water's three phases are ice, water, and steam. Metals have these three phases as well and it is important to know more about these, as these phases are important to 3D printing of metals. A fourth phase of matter, *plasma*, is less well known but quite interesting and relevant to AM as well. High-energy sources such as lasers and electron beams can create plasmas that block or absorb energy used in AM processes. We will get back to that later, but for now we will use the definition provided by *Andrew Zimmerman Jones*,[2]

> Plasma is a distinct phase of matter, separate from the traditional solids, liquids and gases. It is a collection of charged particles that respond strongly and collectively to electromagnetic fields, taking the form of gas-like clouds or ion beams. Since the particles in plasma are electrically charged (generally by being stripped of electrons), it is frequently described as an "ionized gas".

---

[1]Materials Science and Technology Teacher's Workshop metals module Web page, http://matse1.matse.illinois.edu/metals/metals.html, (accessed March 17, 2015).

[2]Andrew Zimmer Jones, "Definition of Plasma", http://physics.about.com/od/glossary/g/plasma.htm, (accessed March 8, 2015).

He goes on to say,

It is odd to consider that plasma is actually the most common phase of matter, especially
since it was the last one discovered. Flame, lightning, interstellar nebulae, stars and even
the empty vastness of space are all examples of the plasma state of matter.

Now you have something to repeat to the plasma physics types should they try to
impress you with their higher knowledge. For the rest of us "condensed matter"
types we will start by building our understanding of metal starting with subatomic
particles and what holds them together.

4.1.2 Elements and Crystals

Everything we see and feel in nature is composed of atoms. We perceive energy
through its interaction with matter. Matter is composed of elements. An element is
composed of atoms, each with the same number of protons in its nucleus and also
contains neutrons and electrons. A metallic element is shiny, conducts heat and
electricity, and can be formed into shapes. A *metallic bond* can be defined as the
strong force that results from sharing of electrons between metallic elements.

Metallic bonding is what makes metals so strong, malleable, conductible, and often shiny
(metallic luster). This strong bond is also why there is such a high boiling point and melting
point for most metals.[3]

A *crystal* is a solid composed of *atoms* and atoms joined together as *molecules*
arranged in a pattern that repeats in three dimensions. A simple arrangement would
be a *cubic structure* with an atom or molecule at each of eight corners of the cube.
A more complex arrangement could be a cube with an extra atom in the middle of
the cube, or one in each face such as shown in Fig. 4.1. In this example, the two
different crystal structures are named "body centered cubic" and "face centered
cubic" corresponding to two different phases, commonly found in alloy steel,
named ferrite and austenite. If an extra atom or atoms of one element is combined
with another pure metallic element we have a new chemical composition called an
alloy. A *grain structure* forms when additional molecules attach to the base crystal
structure to form a larger ordered grouping of crystals, in the same basic orientation.
Grain growth in metals refers to the spontaneous formation or *nucleation* of
multiple grains of varying crystal structures and orientations within a metal object.
When metal is melted, then cooled, and solidified, grains spontaneously form and
grow together to form the *bulk material*. Often when molten metal cools and
solidifies slowly, the grains have more time to grow larger.

Different metal working process produce different grain structures, such as when
casting into a mold, hammer forging, rolled, stretched, or flattened into varying

---

[3]Chem Teacher Web page, "Definition of Metallic Bonds," http://chemteacher.chemeddl.org/
services/chemteacher/index.php?option=com_content&view=article&id=36, (accessed March 8,
2015).

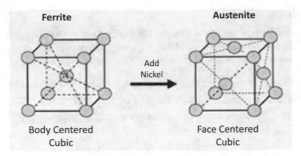

Fig. 4.1 Unit cells of two crystal phases in stainless steel[4]

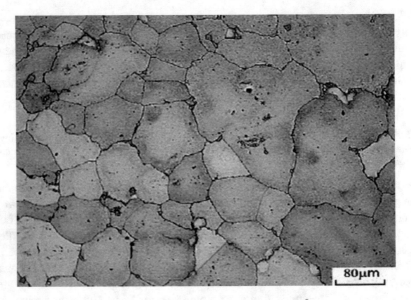

Fig. 4.2 The microstructure microstructureof a cast nickel base alloy[5]

shapes. AM metal parts have their own characteristic grain structure as well dependent upon on the process of melting, cooling, or shaping. These crystal grains may be large enough to view with the naked eye but are more commonly viewed under a microscope. The collection of grains, their size, and orientation is referred to as the *microstructure* of the metal. A cast metal microstructure of a nickel base alloy with a large grain structure, is shown in Fig. 4.2. The size of these grains shown in different colors is about 100 microns or about the diameter of a human

---

[4]Courtesy of IMOA, reproduced with permission.

[5]*Source* Materials Science and Engineering A 487 (2008) 152–161, Effect of various heat treatment conditions on microstructure of cast polycrystalline IN738LC alloy N. El-Bagoury, M. Waly, A. Nofal. Reproduced with permission.

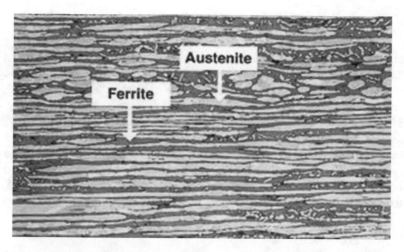

Fig. 4.3 A Roll Formed microstructure in stainless steel Showing Austenite and Ferrite Phases[6]

hair. Metal working processes, such as rolling steel into a flat sheet, can deform these grains into a long flat shape, as shown in Fig. 4.3. Some of the energy used to roll the sheet is locked up into the *cold worked* microstructure, often making it stiffer, stronger, and harder than the cast form. What we have just learned here is the rolling process, or any metal fabrication process, will change the structure of the material, which will change the properties.

It is important to note that these crystals and grains are not perfect. They contain *flaws* and *impurities* that can alter or interrupt the *long-range order* of the microstructure. In metals, impurities within the crystal structure may also be altered by forming and processing. Solid impurities, such as the element carbon, can be left over during the metal ore extraction or refining process or added to improve the strength. Gas impurities, such as hydrogen, may also enter the crystal structure during processing or service and are known as *interstitial elements*. During solidification, *segregation*, or a localized concentration of these impurities can occur and can weaken these regions, such as at *grain boundaries*, possibly leading to failure of the part during service. As an example, during a component failure investigation, one may ask was it cast, forged, rolled, cut, welded, or *3D metal printed*. Additional elements may be added on purpose, by design and are then referred to as *alloy additions*.

Other elements may be added to alter the properties of the bulk material or added later to produce localized chemistry changes such as when applying *surface treatments*. All these matter, as we will return to these concepts along the way as they relate to AM metal processing. Knowing the chemical composition, purity of the alloy, and how it was processed will help you understand the resulting *properties* of the metal and ultimately the performance of the part.

---

[6]Courtesy of IMOA, reproduced with permission.

4.2   Physical Properties

Physical properties are those characteristics that allow us to distinguish one material from the next. For most of us color is often determined by sight, *density* by judging the size versus the weight of an object, strength and hardness by trying to bend or scratch an object. The *thermal and mechanical properties* can vary widely as a function of alloy type and are particularly important as the body of knowledge associated with AM metal parts is very limited. Large databases exist for the material properties and performance of conventional metals compiled over the past 100 years serving the outer space, aerospace, automotive, medical, oil and gas industries. Large bodies of historical knowledge for AM metal processing do not yet exist but are currently in the process of being compiled.

4.2.1   Thermal Properties

Upon heating, metals and metal alloys will reach a temperature where they change from solid to liquid. This can occur at a specific temperature, known as the *melting point* or in the case of metal alloys over a range temperatures referred to as the *melting range*. The melting point or range for many metals can be quite different. As an example, pure lead melts at 328 °C while tungsten melts at 3370 °C. Other melting points for common metals can be found on the Web such as at this link.[7] Continued heating of the liquid metal can reach the boiling point or *vaporization point* (or range). As an example lead vaporizes at 1750 °C and tungsten at 5930 °C. While there are some exceptions, the consolidation of AM metal parts most often requires localized heating of the metal using a laser, electron beam or arc, at or above the melting temperature to fuse or sinter the material. Metal is often vaporized at the location of the heat source impingement upon the molten pool or powder bed. The implications of this melting and vaporization to AM process control are described in detail later in the book.

Thermal expansion (*or contraction*) occurs when metals are heated (grows larger) or cooled (shrinks smaller). The localized heating of a part being built during the AM process imposes a localized heating or thermal variation or *gradient* resulting in differences in temperature across the part. One location of the part is expanding on heating while other portions of the part are contracting on cooling, at the same time. These differences in temperature can lead to distortion or bending of the part and in some cases tearing or cracking of the metal.

Thermal conductivity relates to how rapidly heat energy is dissipated or conducted away within a metal part. In AM, the energy source needs to be adjusted to account for the thermal conductivity of the material to ensure proper fusion of the

---

[7]Engineering Toolbox Web page, "Melting and Boiling Temperatures," http://www.engineeringtoolbox.com/melting-boiling-temperatures-d_392.html, (accessed March 8, 2015).

material. Limitations in AM part design may be encountered, such as minimum wall thickness and must be considered for metals with different thermal conductivity. *Thermal radiance* relates to how rapidly a part can cool down during and after the build cycle such as in an enclosed inert gas chamber, vacuum chamber, or in an open air environment. As we will discuss later in the book, a basic understanding of the thermal properties is useful when designing AM processing conditions.

Viscosity relates to how fluid a molten metal is and how easily the molten pool can be controlled. *Surface tension* is a complex force combining the effect of temperature, chemistry, metallurgy, and physics that makes the top surface of a molten pool the shape of a *bead* and tends to make droplets of water into spheres. Surface tension can have a direct effect on the shape and quality of the AM deposited material. We will mention it again later when talking about chemistry variations in powders and filler materials.

4.2.2 Mechanical Properties

Metal *strength* is the measure of its ability to support a load, such as a compressive (pushing) or tensile (pulling) load. Metal is *elastic* in that it can be stretched and *spring back* to its original shape until it is pulled beyond its *elastic limit* or *yield strength*. A metal sample pulled until it breaks is said to have exceeded its *ultimate tensile strength* (UTS). Steel is commonly understood to be strong, while aluminum is weaker (such as when crushing an aluminum versus steel can).

Ductility is the ability for metal to bend or deform, or conversely, metal lacking ductility can be considered *brittle*. As an example, cast iron is less ductile than mild steel and may break rather than bend under *impact loading*. *Elasticity* or spring back is easily recognized in spring steel but is present in all metals and can have a subtle effect on complex structures. *Elongation* is a measured value that relates to the ability of a metal to stretch beyond its elastic limit until fracture, typically measured against a standard sample dimension. *Toughness or hardness* is a measure of the ability of a metal *to absorb energy, survive impacts, or sustain wear or abrasion* in service. In another example, punches and dies used for hammering or forging in manufacturing operations need the property of high *toughness*, while cutting or drilling tool steel requires high *hardness*. As discussed in greater detail later, the proper selection of AM materials and processes will affect the degree to which the desired material properties can be achieved, repeated, or optimized for a specific application.

Metal *fatigue* occurs as a result of the repeated or cyclical loading and unloading of a part during service and can lead to weakening of the structure. A component may experience low-cycle or high-cycle fatigue depending on the service environment. Fatigue can produce changes to the grain structure or grain boundaries and can result in micro-cracks and built up stresses to the point where large cracks may initiate and propagate, which may result in component failure. Load concentration can be affected by surface conditions such as roughness, notches or geometric features such as sharp inside corners. The surface roughness, layered microstructure, or build orientation of AM fabricated parts may be susceptible to fatigue damage and may require additional finishing operations such as heat treatment, machining, or polishing. However, AM design freedom may allow the avoidance of geometric stress concentration by the avoidance of sharp corners or choice of build orientation and in some cases reducing the risk to fatigue performance.

Other mechanical properties may be affected by AM processing as well. *Fracture toughness* refers to the ability of a material to resist cracking, while *high-temperature creep* refers to the slow and gradual deformation certain materials may exhibit under high mechanical loads or at high operating temperature for extended periods of time. As will be discussed later, the high-temperature creep properties of super alloys may be affected by contaminants picked up during AM processing or within the feed stock used by AM.

In addition to the chemical and physical properties mentioned above, the metallurgical response of a metal is dependent on the process as well. As we will discuss later in the book, the AM process conditions related to "how the metal part was printed" will all come into play. A detailed description of all of the properties of metals and how they are changed by AM processing, is beyond the scope of this book, indeed much is not yet known, but it should suffice to say there will be differences when compared with conventionally processed materials. These differences must be characterized for specific materials and processes. As a reminder, all conventional metal processing operations have their drawbacks and disadvantages, it is just a matter of knowing what they are, how to work around the limitations and how best to optimize the advantages.

4.2.3 Electrical, Magnetic, and Optical Properties

Electrical conductivity and *electrical resistivity* relate to motion or mobility of electrons in or along the surface of a metal as does *optical reflectance and absorption*. Optical properties such as *reflectivity* also relate to the color or metallic luster of a metal as well. A growing number of AM applications, such as 3D printed motors, are beginning to make use of these properties to create functionally hybrid parts as described later.

In addition to part performance, these optical and electrical properties are important to the performance of the AM process itself. As an example, the coupling or reflectance of a laser beam, electron beam or electrical arc, e.g., how well it is absorbed within a bed of metal powder or molten metal pool, relates to the metal electrical and optical properties. We discuss this more in the process specific sections of the book.

4.3 Chemistry and Metallurgy

As with conventional processing, complex *chemical reactions* take place during the AM of metals, some fast and some slow. The speed at which these take place is defined by their *reaction kinetics*. Biocompatibility, as in the implant of a metal part into the body, is one example. Another example of a slow chemical process is *rusting* or corrosion. A common example of a fast reaction is *slag* formation during melting or color change due to oxidation in air during cool down. As we will discuss later, the explosive ignition of a finely divided metal powder is another fast chemical reaction of interest to AM processing. Understanding and controlling chemical reactions gets complicated very fast, but as an introduction you need to know they exist and can differ depending on the metal and processing conditions.

Reactivity of metal alloys often increases with temperature and upon melting. Reactions of metals with oxygen, hydrogen, nitrogen, and moisture are a few of the undesirable chemical reactions that can take place during AM. Conversely, beneficial surface reactions, such as nitriding, may be easily integrated into some AM build sequences.

The degree to which the molten pool requires inert gas shielding and the degree to which these reactions can be tolerated in the finished product, will determine the controls required during processing. As an example, titanium is much more reactive than steel and will require a greater degree of inert shielding both for the molten pool and the part during cool down after the build cycle. In some cases localized shielding such as with an inert nozzle gas may be sufficient (as in some steels) and in other cases a welding grade purity inert gas chamber may be required. If you are processing highly reactive alloys such as niobium or zirconium alloys, only a high purity vacuum environment will be adequate for shielding. Other hazards associated with reactivity can include pyrophoric reactions. Finely divided powders such as magnesium or aluminum can react exothermically, catch fire or create an explosive hazard.

Certain metals and metallic compounds may be toxic, carcinogenic, develop hazardous surface oxides, or generate harmful metallic vapors as a byproduct of processing, such as with the formation of hexavalent chromium vapor when welding stainless steel. It is important to know all the hazards associated with the materials you will be working with. We will mention this more in our safety section, Appendix A.

4.3.1 Physical Metallurgy

Pure metals are rarely used in the fabrication of metal parts, either due to impurities or most often because alloying additions almost always result in either optimized properties or cost benefits. There are many alloys of aluminum that is pure aluminum with additional elements added to change the properties and material performance. In addition there are trace impurities within these alloys that can have a beneficial or adverse effect on these properties as well. The same goes for alloys of steel, stainless steel, titanium, etc., and are often dozens of alloys for each type of material. These alloys are created using combinations of the primary metal type (aluminum, iron, titanium, etc.) and smaller amounts of metallic and nonmetallic elements to change the *chemical composition* of the alloy to obtain specific properties such as strength, ductility, hardness, or corrosion resistance. To put things into perspective, one popular engineering text (Boyer 1994), specific to titanium alloys, provides 1176 pages of material properties, physical and process metallurgy data, for over 50 alloys and process details for a full range of conventional and specialized processes for everything from casting to forming, machining, heating treatment and welding. It is significant to note this relatively new (Boyer 1994) and excellent text makes no mention of what we now call additive manufacturing processes. While the historical data most relevant to AM metallurgy is that provided for weld processing, recent publications provide a good overview of the most recent metallurgical properties of AM titanium (Dutta and Froes 2015) and other common AM processed engineering alloys (Frazier 2014; Herzog et al. 2016; Murr et al. 2012; Sames et al. 2015).

Minor chemistry variations can have significant effects on the bulk properties and performance of a material. Over the history of metal processing, alloy development has closely evolved with processing methods. Many alloys have been developed and either not used due to cost or fallen into disuse due to changing markets or the introduction of better materials. As an example, there is a wide range of aluminum alloys developed specifically for casting and others for *wrought* processing. Tailoring and optimizing alloys for enhanced performance as well as process specific fabrication will undoubtedly undergo the same evolution as AM processing is more widely applied.

Knowledge of time/temperature transformations, describing the evolution of a microstructure upon cycles of heating and cooling over a specific time interval, is critical to controlling the structure and properties of metal. AM has opened up a whole new field of study of material design and is actively being pursued at the university level.

The *weldability* of various metals and alloys can vary significantly and is a good indicator of which metals are best for AM. Therefore, it is important to know what metal alloys you intend to deposit during your AM design phase in order to choose the correct process and testing procedures. It is important to note that most fully dense AM parts, when referred as *sintered metal*, are most often formed as a result of melting and fusion and are most similar in physical metallurgy to metals processed by laser welding. Two good texts providing engineering-level descriptions of the physical metallurgy of welded materials are given by these references: (Easterling 1983; Evans 1997).

4.3.2 Ease of Fabrication

Alloys of steel are often chosen for conventional applications due to strength, availability, cost and ease of fabrication; that is the ability to be cut, bent, drilled, formed, machined, etc. Aluminum alloys are also popular as they are lightweight, easy to machine, and are of relatively low cost. Titanium on the other hand is lightweight, strong, and can resist corrosion but is more difficult to machine, costs more, and is less cost effective to recycle, therefore limiting its usage. But, as will be discussed in more detail later, AM can change the equation regarding the ease and cost effectiveness of fabrication when building complex metal parts. Titanium is one of the most popular, easy to print AM metals and despite the high cost of AM titanium powder, it can provide significant cost savings when fabricating complex, large or high-value parts where conventional processing methods produce significant amounts of scrap and waste material. Buy-to-fly ratio is a term used in aerospace to compare the volume of raw starting material to the volume of material within the final part. As an example, conventional processes, such as machining of large complex aerospace part, may have a buy-to-fly ratio of 40:1 when compared to an AM process of 4:1. In comparison to titanium, aluminum alloy parts can be more difficult to AM fabricate, with a less attractive buy-to-fly ratio saving, while AM parts made from steel may not be cost effective when compared to conventional processing methods.

4.3.3 Process Metallurgy

Most folks have heard or read about *tempered* steel, *forged* steel, or *case hardened* steel because it is printed on the side of their knife blade or crescent wrench. They also know of the process of casting because the *cast iron* gate at the historical museum is made of this and *wrought iron* appears to have been hammered or worked. They may also know welding is used to construct steel frames of buildings and welds are things that can break during an earthquake, but that is about all they know about the useful properties resulting from the process used to fabricate a metal object.

It is important to elaborate on a few of these metal processing examples to introduce a few references and emphasize that *how you process the material will affect how the part will perform. The process drives the metallurgy, the metallurgy drives the properties, and the properties will affect the performance.* But, as you will see later in the book, there are some advocates for AM processing that want to turn this around and drive the metallurgy by changing the process, in real time, while building the part. They ask, why heat-treat the entire part when only sections of it need heat treatment? Why case-harden an entire part when only certain regions need it? Why not use different materials in specific, such as depositing wear-resistant material only where you need it? In many cases the answer is, because it is too complicated to be cost effective for conventional methods. In certain cases AM metal processes may be able to accommodate this added complexity but in other cases not. It is easier to think outside of the box than to make money outside of the box. However, in some cases AM technology, which was only being talked about and studied in universities ten years ago, has grown into viable business models today. For now, let us build a foundation and stick to the basics.

An introductory course in metallurgy or a good metallurgical text (Boyer and Gall 1985) can go a long way to introduce the subject, but to start we will mention a few examples relevant to AM. The most relevant process metallurgy texts are those that relate to laser, electron beam or arc welding as these processes use the same types of high energy sources for melting (Lienert 2011; O'Brien 2007). The open literature can provide excellent information from the most recent academic sources. Use the search terms provided in this book to search the Web and many of the references provided in the bibliography to access additional open source information. Here is one example (Gong 2013) of an excellent Ph.D. thesis providing a scholarly study of the generation and detection of defects in metallic parts fabricated by selective laser melting, electron beam melting and their effects on mechanical properties. Another excellent link to open-source work is the archive of the International Solid Freeform Fabrication Symposium,[8] hosted each year by the Laboratory for Freeform Fabrication and the University of Texas Austin.

An important differentiation in AM metal processes is that of *sintered* versus *solidification* microstructures. In the early days of rapid prototyping, starting the late 1990s some of the technology being used to form metals parts did not fully melt the metal creating a weaker part by the process of sintering.

4.3.4   Sintered Microstructures

What is metal sintering? Makers will often encounter references to *sintered metal* shapes and laser sintered processes. In the classic definition, sintering of metals involves joining without complete melting, coalescence or solidification. It may be referred to as partial fusion. In sintering, molecular bonding is primarily accomplished through *diffusion*. Diffusion in metals is a solid state process where atoms and molecules can move about and in the case of sintering grow together by forming metallic bonds often at elevated temperature and extended time intervals.

Sintered microstructures are not always 100% dense and may require hot isostatic pressing (HIP) processing to fully consolidate the material. In an effort to attain a near 100% fully dense deposit, many selective laser sintering (SLS) processes, as applied to metal, have evolved to become selective laser melting (SLM) processes. We will discuss this in detail later in the book.

---

[8]Archive of the Laboratory for Freeform Fabrication and the University of Texas Austin, http:// sffsymposium.engr.utexas.edu/archive, (accessed March 15, 2015).

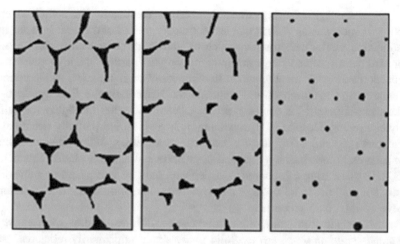

Fig. 4.4 Three stages of powder sintering[9]

Sintering is a material consolidation process, often applied to ceramics and metals that use heat, often pressure and partial melting, to break up surface oxides and bring powder particles close enough to allow diffusion and grain growth creating a bond between the particles. Conventional powder metallurgy processes have evolved over the centuries as a viable and cost effective means to form functional shapes without complete melting. While cost effective for certain applications, these processes do have both limitations and advantages. Figure 4.4 shows three stages of powder sintering dependent upon the material, and/or increased temperature, pressure and time, These three stages, left to right, show an increase in the density of the material and a decrease in the void volume or un-sintered spaces between particles.

Conventional powder metallurgy processes often rely on coating a powder with a binder and the use of molds and pressure to form a *green shape*. The binders are burned off in a furnace and the remaining metal powder sintered by use of high temperature, relying on diffusion to join the particles into a final net shape to attain functional metal properties. Early rapid prototyping methods relied on plastic prototyping equipment and coated metal powders to build up a 3D shape, without the need for molds to form a *green part*, which was then *burned out* and *infiltrated* by an additional heating and brazing step. This infiltration uses a lower melting point metal, such as copper or bronze, to be melted, drawn into the porous part and solidified to create a solid composite metal object. Current powder bed fusion processes, relying on sintering alone, without binder or infiltration steps, can produce deposits with a higher degree of density, achieving function properties without the need for infiltration.

---

[9]©EPMA—www.epma.com, reproduced with permission.

A big advantage of using sintered metal processing is the ability to create functional objects using metals that are not easily melted and fused. Powder metal consolidation works well for parts such as hard high carbon alloy steel gears or other hard metals, often using a nonspherical powder particle shape to assist in cold compaction and mechanical bonding. In other cases, high pressures and temperature cycles are used to compact and improve the density of the final product. Hot isostatic pressing (HIP) is one such process, but it also relies on highly specialized, costly equipment. Despite HIP processing, micro voids and partially fused particle boundaries may still exist, adversely affecting properties, although some applications such as bronze bushings depend on a porous structure to hold oil and lubricants. HIP processing of sintered AM deposits may be performed to achieve full density but may not in all cases prevent reduced elongation at fracture or failure to achieve optimal bulk properties.

For AM applications, sintering can be further differentiated as *solid phase sintering* and *liquid phase sintering*. Solid phase sintering primarily relies on heating and diffusion to "neck" and join particles together, often resulting in a porous structure, while liquid phase sintering heats metals to the point where some of the alloying constituents melt and further assist the bonding process. AM sintered metal can have densities as low or high as 94–99+%.

Selective metal laser sintering technology, using powder bed fusion technology, has evolved to achieve near 100% dense deposits in many materials. This blurs the boundary between classical sintered microstructures, often with less than full density and relying on diffusion only, to parts displaying a fully evolved solidification microstructure, grain orientation and phase evolution resulting from a traveling melt pool.

Figure 4.5 shows laser-deposited material produced under varying conditions of scan speed and hatch spacing resulting in varying degrees of porosity resulting from a range in processing parameters resulting in vary degrees of penetration. Magnified views of the cross section of 6 samples deposited were photographed in the as-polished condition to reveal varying degrees of porosity due to melting conditions. These results range from a sintered microstructure with voids and porosity, to a melted microstructure achieving near full density (Vandenbroucke 2007).

4.3.5   Solidification Microstructures

What is melting and fusion? Metal fusion occurs with melting and overcoming any surface conditions at the melted interface to allow coalescence and mixing, followed by cooling, and solidification. Solid-state molecular bonding can also occur in metals if you get the material surfaces clean enough and the molecules close enough, but melting makes this easier. We will revisit solid-state bonding later when talking about diffusion and ultrasonic consolidation. However, most AM metal processes rely on fusion. *Melting, coalescence, and solidification* allow the benefits of a 100% dense deposit and a continuously evolved microstructure with

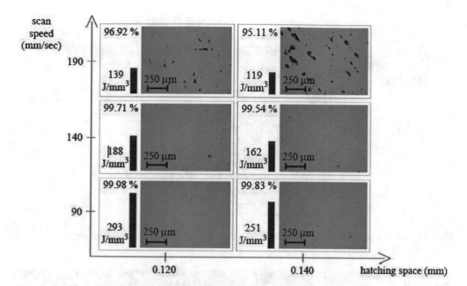

Fig. 4.5 Voids in an AM *Metal Deposit Resulting From Various* energy density Parameters[10]

the associated benefits of improved mechanical properties and performance. Casting is a metal fusion process, but it does require a mold and therefore is not considered an additive manufacturing process.

AM metal microstructures created by melting and fusion are most closely related to multi-pass weld microstructures and those obtained from laser and EB weld processing of metals. Figure 4.6 show a schematic of the microstructure of a steel *casting* with large columnar grains oriented inward as following the solidification path toward a central region of *equiaxed* grains. The cast microstructure may be compared to that of the microstructure created by a single melt pass laser weld in 304 stainless steel, shown in the weld cross section Fig. 4.7. The melted penetration weld shape is referred to as the weld fusion zone. Also note that the small grained microstructure of the *fusion zone (FZ)* microstructure is significantly different from the large grains present in the base metal. We will talk more about this later as it is important to AM metal parts because as-deposited AM materials are essentially made up of an all-welded microstructure. Figure 4.8 (Fulcher 2014) shows a cross section of AM microstructure of an aluminum alloy revealing the multiple buildup of successive deposition paths used to form the part. The bright colors are results of an electrolytic etch and viewing the sample under a microscope using polarized light. Figure 4.9 also shows the structure of the multiple melt paths and layers in

[10]*Source* Ben Vandenbroucke, Jean-Pierre Kruth, "Selective Laser Melting of Biocompatible Metals for Rapid Manufacturing of Medical Parts," SFF Symposium Proceedings, D.L. Bourell, et al., eds., Austin TX (2006) pp. 148–159. Reproduced with permission.

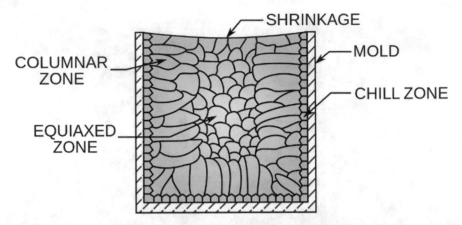

Fig. 4.6 Illustration of the macrostructure of a cast ingot. "Structure cristalline lingot,"[11]

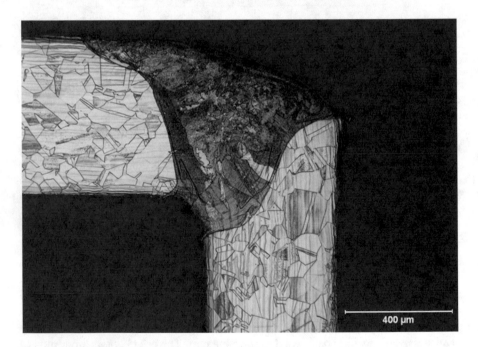

Fig. 4.7 Laserweld cross section showing a refined fusion zone microstructure

SLM deposited aluminum. Note the change in the deposition melt path and deposited pattern as the laser is scanned along the outer edge of the deposited shape, shown toward the right of the picture.

---

[11]Courtesy of Christophe Dang Ngoc Chan under CC BY-SA 3.0: https://creativecommons.org/licenses/by-sa/3.0/, https://commons.wikimedia.org/wiki/File:Structure_cristalline_lingot.svg.

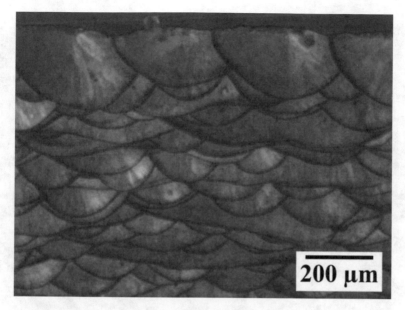

Fig. 4.8 Optical micrograph cross section of as-built AlSi10 Mg, with electrolytic etch[12]

The highly repetitive order of an AM deposit can have repetitive and directionally dependent effects upon the bulk material. Rapid solidification rates of 10^3–10^5 °C/s can produce highly refined microstructures, suppress diffusion controlled solid state phase transformations and form *nonequilibrium* or *meta-stable phases*. Note that the microstructure shown in Fig. 4.9 is but one example, as vendors have developed sophisticated processing methods to use to advantage, the ordering of repeated *or tailored microstructures* using various deposition or scan paths and processing schemes.

Microstructural evolution is most strongly related to peak temperatures, time at temperature, and cooling rates, although impurities can also play a part. The effect of cooling rate on grain size, segregation of alloy constituents and phase fraction can greatly affect the properties of the bulk deposit. *Reaction kinetics*, or the rate at which chemical or metallurgical reactions occur, is also an influence regarding the crystallographic phase structure and resulting properties. In a familiar example, certain alloys of steel may be quenched, such as that of a sword, to make it harder and stronger by favoring the growth and retention of the harder phase *martensite*, over other softer phases. Many of the solidification microstructures created in an AM deposit are those typical to laser or electron beam welds or multilayer laser

[12]*Source* Benjamin A. Fulcher, David K. Leigh, Trevor J. Watt, "Comparison OF ALSI10MG and AL 6061 Processed Through DMLS," SFF Symposium Proceedings, D.L. Bourell et al., eds., Austin TX (2014) pp. 404–419. Reproduced with permission.

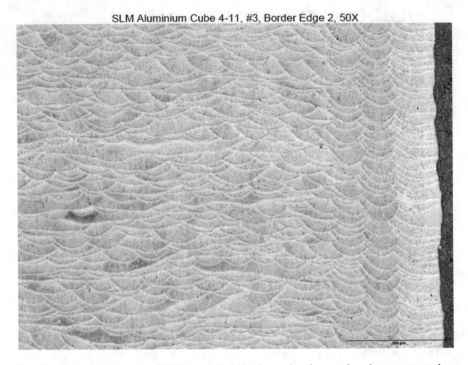

Fig. 4.9 Optical micrograph of SLM deposited aluminum showing an altered scan pattern along the *right* edge of the deposit[13]

cladding. Researching the metallurgy related to high-energy beam welding can provide a relevant body of knowledge to study when exploring the potential uses of a new AM processes and material.

Epitaxial growth describes the continuation of an existing crystalline structure from either a single crystal or from existing preferentially orientated grains into a modified structure with similar orientation. In AM processing, the melt back from the bead being deposited into a previous layer can solidify and continue the previous layers' grain orientation and preferentially propagate the properties of that existing structure. This can be advantageous or detrimental. Many AM process procedures randomize or change the deposition path and resultant solidification conditions to avoid this preferential bias of properties in order to minimize stress concentrations, warping or other *anisotropic* properties within the bulk deposit. In other cases, researchers are attempting to predict or harness this level of microstructural control, often referred to as microstructural engineering (Murr et al. 2012).

---

[13]Courtesy of SLM Solutions N.A. Inc., reproduced with permission.

Microstructural defects such as segregation of impurities, hot cracking, exaggerated grain growth or undesirable phase formation can result from thermo-mechanical conditions unique to AM processing. A wide range of conditions can result from rapid cooling of laser deposited metal, slow cooling, or long dwell times at elevated temperatures in partially impure atmospheres. In addition, losses of alloying elements due to heat source-induced vaporization are but a few of the known unique conditions of these new and evolving AM technologies.

Forging is another process that uses pressure and deformation to create a shape with a microstructure that is stronger due to *work hardening*. In certain cases near net-shaped objects produced by AM may be used as forging preforms, utilizing forging as a post-processing step to achieve the desired microstructure, work hardened properties, and final shape.

Heat treatments (HT) such as *annealing, solutionizing, homogenization, recrystallization, or precipitation hardening* may also be used in addition to the metal forming process to achieve the desired or necessary properties in an AM metal component. Heat treatments provide a high-temperature heating cycle below the melting point for an extended duration, allowing grain growth, softening, relaxation of stresses, or to create a uniform microstructure. Heat treatments may be performed in high purity inert or vacuum chambers. Heat treatment durations may require hours to allow the entire part to reach the desired temperature (*soak time*) and remain at temperature long enough to allow the desired metallurgical changes to occur. In the case of heat treatment for AM parts, modifications to typical heating treatment schedules may be needed to accommodate complex part geometry, or the characteristic highly refined or banded AM microstructures resulting from the layer by layer method of fabrication.

Hot isostatic pressing (HIP) is a specialty process, using specialized equipment, involving high temperature, high pressures and time to consolidate powder, close voids, or heal defects within a material. Post processing of 3D metal parts often uses HIP to modify the as-deposited AM microstructures to attain the desired metal properties and ultimately the desired part performance. HIP post processing can reduce or eliminate the directional dependent of anisotropic properties of as-deposited AM material resulting in bulk properties in some cases that are equivalent of better than those of wrought material. HIP equipment and processing can be costly and impose size limitations upon candidate components. In cases such as AM medical parts, the small size allows batch HIP processing and can be cost effective. Greater detail will be provided regarding HIP processing of AM parts later in the book.

Tempering is a heat treatment used to reduce hardness and improve toughness. Quench hardening performs a rapid cooling to harden metals by locking in or retaining desirable phases into the crystal structure. These processes are often performed to change the microstructures of an as-processed metal part to achieve more uniform or more desirable properties. Case hardening is a heat treatment that will change the surface chemistry of the outer layer of metal often by the addition of carbon or nitrogen (such as by *carburizing or nitriding*). Although not commonly applied during AM, these are examples of metal processing techniques that impart properties that may be achievable during the AM printing of a part.

It is worth pointing out that the control and optimization of the many metallurgical effects associated with 3D metal printing are yet to be understood for a wide range of engineering alloys and new materials. The freedom and flexibility of AM processing allows access and modification to all locations within the part as it is being built, but the complexity of all possible combinations of materials and processes to achieve location-specific properties are far beyond current engineering competencies. In most cases, the highest degree of deposit quality can be attained by following the guidelines and procedures developed by using vendor supplied materials and procedures. We will touch upon these issues and provide references later in the book.

There are other ways to achieve 100% density that do not use melting at all. Pressure and deformation, often combined with heat, may be used to bring surfaces into intimate contact and break up layers of surface impurities and achieve a forged component. Ultrasonic consolidation is another AM process variation that does not require melting. Conventional metal casting will be mentioned later when describing hybrid AM-based methods that utilize 3D printing to create wax patterns or molds and conventional casting to form the metal part. As a reminder, while mentioning other processes that use 3D printing that result in a metal part, this book will focus on fused metal deposition using melting or sintering processes and localized heat sources.

4.3.6 Bulk Properties

To reward those readers who stuck it out to get to this portion of the chapter, let me remind you that these properties, as measured for a specific solid material, are generally referred to as *bulk properties*. That is, they are for the most part uniform throughout the object and do not depend on the size or shape of the part. A part with a homogenous material or structure will generally display these properties uniformly throughout. As an example, a part may exhibit uniform density and hardness, deviating only at isolated flaw locations. AM processes based on layer-by-layer deposition can display *anisotropic* properties that vary depending on the orientation of the part with respect to the build direction, orientation, and AM build parameters. The properties in the vertical build direction such as strength or ductility may be different in another orientation along the plane of a given layer. If porosity or voids are present in a less than fully dense object, the bulk properties will deviate from those measured by a standard test specimens. We will discuss this in detail later but for now, as a take away message, the definition and determination of AM bulk properties may differ from those of other common engineering materials, requiring care in the comparison between conventional and AM-processed materials, both in the as-deposited and heat treated or HIP condition.

Uniform and repeatable bulk properties are one of the biggest challenges facing AM-deposited metals. It is important to note that those challenges exist for all metals but conventional alloys and processes have existing databases to reference. To appreciate that difference, one can refer to the MMPDS Handbook, The Metallic Materials Properties Development and Standardization handbook[14] and its predecessor MIL-HDBK-5, as they are the preeminent sources for aerospace component design. MMPDS is a unique, cost-effective collaboration between government agencies and industry which establishes allowable properties for aerospace metallic materials and fasteners. Significant efforts are underway at corporate research labs to develop materials properties databases for AM-fabricated materials.

4.4 Forms of Metal

4.4.1 Commercial Shapes

Metal comes in many forms, the most common of these being commercial shapes and fabricated or manufactured products. We are all familiar with commercial metal shapes such as sheets, plates, pipes, tubing, angle iron, the I-beams used to construct buildings and with commercial metal products ranging from kitchen products to automobile engines. Mass production of common metals and shapes such as those made from steel and aluminum has reduced the cost of these products to a minimum. Common metals are available in a wide range of commercial shapes and thickness while shapes of less common metals are more limited. AM metal machines rely on specialty metal powders and wire that may be difficult to obtain for specialty alloys. Some of the more common metal types and applications are described below. A limited selection of these alloys is being optimized in powder form for AM processing, while other already exist in wire form.

Aluminum is a lightweight common engineering metal with good electrical conductivity, mechanical and thermal properties, is available at a relatively low cost and is easily machined or welded. Aluminum alloys may be heat treated and homogenized to obtain a uniform range of properties. It is used in prototyping, aerospace, automotive, consumer products, and short-run production applications.

Maraging steel and tool steel (martensitic age hardened steel) is used in injection and die casting molds and tooling to produce strong and robust parts for longer run

---

[14]The Metallic Materials Properties Development and Standardization handbook, http://projects.battelle.org/mmpds/, (accessed April 10, 2015).

production applications. Conformal cooling may be integrated in molds to improve molding process performance. Post processing of molds using machining, EDM, heat treatment and polishing is used to produce high quality durable surfaces.

Stainless steel alloys offer strength, chemical, and corrosion resistance at a cost less than super alloys such as cobalt-chrome, Co-Cr. They are often used in a wide range of applications such as medical devices, food processing, pharmaceutical, aerospace, and marine products. They offer good fabricability and are easily deposited using AM methods. Alloy 17-4 PH can be precipitation hardened enabling use in short production run tooling.

Nickel-based alloys such as Inconel 625 and 718 are nickel chromium super alloys exhibiting good creep and tensile strengths at elevated temperatures, for long periods of service. They are widely used in aerospace applications such as aero engine and combustion chambers. Corrosion resistance makes them attractive as a cladding material for use in seawater or other harsh environments such as in the oil, gas, and chemical industries. They may also be used in prototypes and offer good machining and weldability.

Titanium alloys offer excellent strength, corrosion resistance, biocompatibility, low thermal expansion and light weight. The most common alloy Ti-6-4 (also Ti-6AL-4V) combines pure titanium with 6% aluminum and 4% vanadium. The ELI (extra low interstitial) specifies a higher degree of purity for critical performance applications. It is often used in aerospace and medical implant applications where cost is secondary to in-service performance or biocompatibility. They can be machined and welded although with less ease than aluminum alloys. The high cost of these specialty powders limits their use in consumer products, but with the high utilization and reduced scrap rate, AM processing can become attractive for the fabrication of larger aerospace components.

Cobalt-Chrome alloys are high-strength super alloys offering excellent high temperature performance, toughness, and corrosion resistance. They offer bio-compatibility and find uses in medical and dental applications. Mechanical strength at temperature and hardness make them attractive for aerospace applications. Heat treatments such as stress relief or hot isostatic pressing may be required to achieve the desired properties. Less easily machined than other metals, these alloys are often cast into shapes making them good candidates for high-value AM parts.

Refractory metals such as tungsten, molybdenum, rhenium, niobium, and tantalum offer extreme temperature structural performance and very high density. They are hard to fabricate into complex shapes using conventional processing such as rolling, machining, or welding; therefore AM produced shapes may greatly extend their application. Advanced applications such as reactor and furnace hardware are possible, while beam collimators and anti-scattering grids have been demonstrated for lead-free medical imaging applications using tungsten. As will be described later in the book, 3D printing using binder jet technology may be used to form difficult to fabricate metal shapes without melting if full density of the deposit is not a requirement.[15]

Other commercial powders such as those for copper alloys, bronze, precious metals such as gold and platinum and other materials are being used to build AM hardware. Low melting point metals are easily cast using conventional methods such as their use in jewelry. In these cases, 3D printing may simply be used to create a wax pattern or a mold to facilitate the use of conventional casting to realize the final metal shape. Hard materials such as tungsten carbide (WC), titanium carbide (TiC), or titanium nitride (TiN) are often used as cladding and applied by blown powder methods. Nickel titanium (NiTi) a specialty alloy developed at the US Naval Ordinance Laboratory is known as *Nitinol* (NiTiNOL) and is of the class of shape-memory alloys (SMAs) known for known for the ability for large changes in deformed shape as a function of temperature and the ability to fully recover the original shape as a function of time and temperature. Nitinol also features bio-compatibility allows use in applications such as medical devices. As with other specialty materials it can be difficult to process such as by melting and machining which has limited its use commercially. *Shape setting*, heat treatments, and maintaining purity during processing must be carefully controlled. AM processing offers an opportunity to reevaluate the use of these specialty materials and in some cases find new commercially viable end products and high performance components.

Other specialty materials are being optimized in powder form AM application but are currently available only in limited amounts as specialty orders.

4.4.2 Metal Powder

Powder Metallurgy technology is a field unto itself and has been successfully applied for many years to process metals that do not lend themselves to casting or wrought processing. It can also be cost effective for relatively simple, smaller parts fabricated in large lot sizes. Gears are but one application where hard durable metal can be pressed and sintered in durable commercial shapes. Depending on the metal or alloy, commercially available powder comes in many sizes and shapes and is produced using a wide range of processes.

Metal powders for fused or sintered deposition have been widely used in cladding, flame or plasma spray applications. However, the requirements for metal powders for AM are more stringent as the size, shape, and chemistry are critical for a successful and repeatable process. Conventional metal powders such as those

---

[15]HC Starck Web site for refractory materials, http://www.hcstarck.com/additive_manufacturing_w_mo_ta_nb_re/, (accessed March 21, 2015).

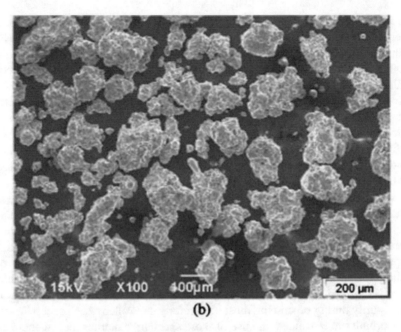

(b)

Fig. 4.10 Iron-based water atomized powder[16]

produced by water atomization, may be angular (Fig. 4.10), irregular, or agglom-
erated, in shape with sizes ranging widely from submicron dimensions to well over
100 microns. Power shapes such as these are unsuitable for AM PBF processes.
Powders currently used for AM most commonly range from 10 to 105 microns in
size and are generally spherical in shape, allowing powder bed machines to spread
fine layers of powders evenly and powder feed systems to deliver an inert gas fed
stream of powder smoothly without nozzle clogging. The hydride–dehydride
(HDH) process uses a hydrogen reaction to form and extract a metal hydride
compound which is later reacted back to metallic form. Figure 4.11a shows HDH
Ti64 powder that would need and additional plasma melting process to produce a
spherical shape particle.

[16]*Source* Effects of annealingannealing on high velocity compaction behavior and mechanical
properties of iron-base PM alloy, Hongzhou Zhang, Lin Zhang, Guoqiang Dong, Zhiwei Liu,
Mingli Qin, Xuanhui Qu, Yuanzhi Lü, Powder Technology 288, (2016), 435–440, DOI: 10.1016/j.
powtec.2015.10.040. Reproduced with permission.

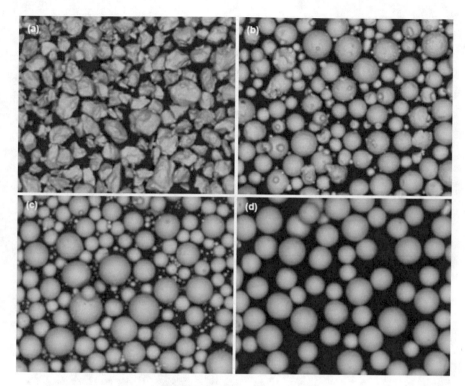

Fig. 4.11 Powder produced for AM processing by **a** HDH, **b** gas atomization, **c** plasma atomization, and **d** plasma rotating electrode processes[17]

Titanium alloy Ti-6-4 powder produced by the HDH, gas atomization, plasma atomization, and the plasma rotating electrode processes and is shown in Fig. 4.11a–d.[18] Scanning electron microscope images of these spherical metal powders show differences in the particle size distribution (PSD), that is the range of particle sizes present in a single batch of powder.

The size, shape, and chemical purity is different for each process and alloy and may be optimized for laser and electron beam processes, powder bed or gas stream delivered. Powder sieving may be required to assure the optimal range of powder dimensions and to remove irregular necked, joined or satellite powder particles. Not all commercial metal powder will work for AM although some systems are more

[17]*Source* Introduction to the Additive Manufacturing Powder Metallurgy Supply Chain, Johnson Matthey Technol. Rev., 2015, 59, (3), 243. Reproduced with permission.

[18]Introduction to the Additive Manufacturing Powder Metallurgy Supply Chain, Jason Dawes, Robert Bowerman and Ross Trepleton, Johnson Matthey Technol. Rev., 2015, 59, (3), 243, http://www.technology.matthey.com/article/59/3/243-256, (accessed April 19, 2016).

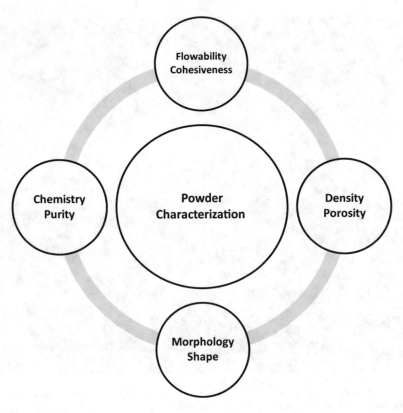

Fig. 4.12 Properties related to powder characterization

tolerant than others to powder shape, size and purity. LPW Technology provides a Web link[19] to technical information of powder production with graphics illustrating each process. The graphic in Fig. 4.12 shows various properties related to powder characterization.

These methods produce a wide range of powder shapes and purities at a wide range of costs. Each method is best suited for specific materials, AM processes, and applications. Figure 4.13 shows a typical gas atomization process where the metal alloy is induction or furnace melted and atomized with a stream of gas, typically argon, collected and sieved into the desired range of sizes. The gas atomization process can produce high purity, spherical powders, but at an added expense, gas porosity entrapment within atomized powder may also be an issue and in some cases has been proposed as a source of gas porosity within the as-deposited AM

---

[19]LPW Technology provides a Web link to technical information of powder production with good graphics illustrating each process, http://www.lpwtechnology.com/technical-information/powder-production/, (accessed March 13, 2015).

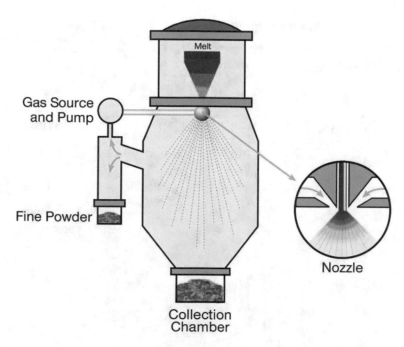

Fig. 4.13 Gas Atomization Process[20]

material. Other impurities such as iron can be picked up during the powder creation and is considered undesirable for certain applications. Plasma atomization can produce very spherical uniform powders requiring a wire form of feedstock, therefore increasing the cost and decreasing the range of powder types. Figure 4.14 shows the Advanced Plasma Atomization™ process Of AP&C, where metal wire is fed into the atomization chamber and melted by plasma heat sources to produce highly spherical powder. Titanium powder produced by this process in the size range of 106 microns is shown in Fig. 4.15.

Commercial equipment vendors are supplying proprietary metal powders for use in their machines, but at a cost premium to reflect their assurance the powder is suitable for use with their AM machines. The current supplies of commercial powders available for AM processes will increase in quality and decrease in cost responding to the market needs and expanding the range of applications. We discuss this in more detail later in the book.

Powder optimization for AM is needed to address a number of current limitations. As stated above, low cost, clean and uniform powders must be readily available from commercial sources. Flow characteristics, size, and particle shape

---

[20]Courtesy of LPW Technology, reproduced with permission.

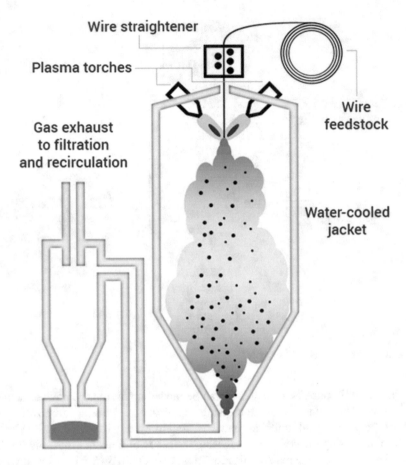

Fig. 4.14 AP&C Advanced Plasma Atomization Process[21]

must be tailored to enhance powder delivery such as spherical powders. Commercial powders produced by precipitation or crushing may need to be reprocessed to impart the desired characteristics, such as by remelting and sieving to attain the proper size and shape. Powder chemistry and alloy composition may need to be altered to ensure alloying constituent concentrations are maintained to accommodate any preferential vaporization or loss during the process. Research efforts are underway to understand and develop methods for the characterization of AM powder properties leading to improved test procedures leading to the

---

[21]Courtesy of AP&C Advanced Powders and Coatings, reproduced with permission.

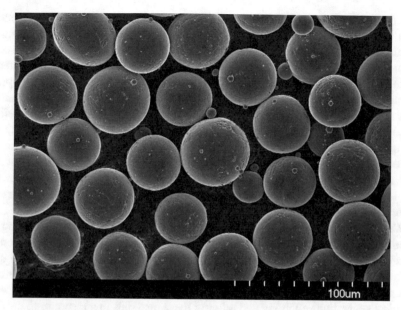

Fig. 4.15 Titanium powder produced by the APA process[22]

development of standards for AM powders and the parts produced from these powders.[23] Ongoing research continues to search for lower cost, more environmental friendly ways to extract metals and produce high quality powders for AM.[24]

4.4.3 Wire and Electrodes

Metal wire is commercially available in many forms and in many alloys although much of it has not been optimized for use in welding or melting applications, such as AM. Weld wire is manufactured with close control of chemistry, impurities and dimensional control. Alloy filler wire for commercial arc welding is used in a wide range of applications and has been developed and optimized over the past half century for specific materials and specific ranges of parameters, such as current and

---

[22]Courtesy of AP&C Advanced Powders and Coatings, reproduced with permission.

[23]April L. Cooke; John A. Slotwinski; Properties of Metal Powders for Additive Manufacturing: A Review of the State of the Art of Metal Powder Property Testing, NIST Interagency/Internal Report (NISTIR) 7873, http://www.nist.gov/manuscript-publication-search.cfm?pub_id=911339, (accessed March 13, 2015).

[24]One such example is provided by the company Metalysis, Rotherham UK is working with Sheffield Mercury Center is developing a cheaper way to create titanium powder using electrolysis rather than the Kroll Process promising 75% cheaper AM titanium and tantalum powders., http://www.metalysis.com/, (accessed March, 13, 2015).

travel speed. Filler wires include a wide range of solid metal in straight or spooled forms. Hollow "cored" filler wire wraps flux or alloying powder in a sheath of metal, providing an economic means to tailor the chemistry of the final welded deposit but as of yet has not found applications in AM processing. These materials range in cost, form, and quality with a wide range of diameters and impurity content. They have been certified for use with laser and electron beam welding applications and are routinely specified as part of any formal standard welding procedure or code. The filler is often supplied as spooled wire and fed directly into the molten pool to achieve melting. As AM applications increase and material property databases become populated, commercial weld wire alloys used for AM may need additional chemistry adjustments to accommodate build conditions typical to new AM heat sources or deposition schedules and the time/temperature history of the part during the AM process.

Welding *consumable electrodes* carry weld current as part of the welding arc circuit functioning both as an electrode and filler. The wire electrode melts and is transferred as liquid metal across the arc into the molten pool. Various spool sizes and electrode diameters accommodate various deposition rates and effective bead dimensions. Manufacturers of wire and electrode filler metal products have spent technical generations evolving and refining these materials in a response to need for new processes and weld deposit properties. The welding consumable industry may spend another few decades following AM process trends and developing new alloys. The welding equipment and consumable suppliers' Web pages are a good place to start to gain more information regarding welding fillers and wire electrodes. A few of the main vendors are listed at the end of this book in the AM Machine and Service Resource Links section under the Welding Equipment and Consumable Suppliers.

These filler materials have been used in applications of weld cladding and shape weld build up and have now been demonstrated in use by AM metal systems. The wide range of alloys available, existing base of application knowledge and performance data give these materials an advantage over the more limited range of powder alloys certified for critical use in AM applications. Disadvantages when using wire filler include the need for a larger molten pool to melt the wire, which reduces deposition accuracy, increasing heat input, and may result in increased distortion but with the benefit of higher deposition rates. Advantages to wire filler include ease of use when compared to powder handling. Specialty metal wire not specifically manufactured for welding applications may be used in AM processing, but should be carefully evaluated when considered for AM use. These considerations will be discussed in greater detail later in the book.

4.4.4 Graded Materials

3D metal printing may provide the opportunity to change material or material properties as a function of location within the part. By changing the material chemistry or deposition parameters, the properties of the deposited material can be changed, either intentionally or as a result of a process disturbance. AM may afford us the opportunity to change, grade, or tailor these properties within the part creating a functionally hybrid component. The potential exists to grade any of the bulk properties described above by changing the materials, design, or processing parameters. We discuss this in greater detail later in the book. Graded metal compositions are well known to metallurgists associated with weld cladding or surface treatments. Terms such as *base metal dilution, incomplete mixing,* and *segregation of alloying constituents* are well known by the welding metallurgist and are understood as ways to functionally grade a cladding or produce a transition joint between dissimilar base metals. The higher degree of spatial and time/temperature control offered by AM holds promise to advance graded material processing to a new level.

Graded functional properties may be extended over domains other than spatial. Shape-memory alloys (SMA), also known as smart metal, memory metal, or memory alloy, are typically alloys of copper aluminum, copper nickel, or nickel titanium. They can change phase and shape during heating and cooling and are being investigated with respect to the unique benefits of AM processing (Hamilton et al. 2015).

Bimetallic combination may be fabricated to actuate switches while graded materials with differing thermal properties may integrate features used for strength with features and materials optimized for heat conduction or low thermal expansion. The possibilities for AM to combine different materials and functional features open a whole new world of design possibility.

4.4.5 Composites, Intermetallic, and Metallic Glass

Intermetallic materials that deviate from the classic definition of metals, are being used where high temperature and high-strength performance is required for advanced applications. Metal matrix composites are materials in which nonmetallic compounds are added to a base metal to impart a combination of properties not attainable in a single material. An example would be tungsten carbide or hard metal particles embedded in a soft, tough, ductile metal matrix for cutting tools or abrasive features; another is an aluminum matrix with silicon carbide filler to provide enhanced strength while retaining low weight. AM processing can extend the application of hard materials to complex or difficult to form features not achievable by conventional processing or by varying the strength or thermal performance of features by changing the deposit to a metal matrix composition.

Metal-filled composites could locally impart a desirable property specific to location within the bulk part offering additional benefits by lowering material costs or increasing the ease of fabrication. In one example, polymer-coated tungsten powder may be fused using plastic 3D printing technology to form complex radiation collimator for medical imaging. Chapter 10 gives us greater insight into this type of out-of-the box thinking.

Intermetallic materials are combinations of metallic materials often forming large complex crystal structures that can display unique properties within bulk materials, often as coatings or strengthening phases in super alloys. They often deviate from the common crystal structures and definitions of more common metals. Nickel aluminide or titanium aluminide (TiAl) are two such examples most commonly found as strengthening phases within aerospace super alloys but also being investigated for the fabrication of aero engine components using. TiAl has low density, high oxidation resistance, high Young's modulus (a measure of elastic performance) and can be used at much higher service temperatures (Murr et al. 2010). These intermetallic materials may also display unique chemical and physical properties, such as uses in hydrogen storage, in superconductivity and as magnetic materials. Amorphous materials and metallic glasses are also types of intermetallic materials where rapid cooling has prevented the formation of the typical crystalline structure of metals. Amorphous metals, often referred to as metallic glasses are metallic materials that do not have the typical crystalline structure of common metals. They are often produced by rapid cooling and can display properties beyond those of conventional metals such as high strength, low Young's modulus, and high elasticity. Research show that cooling rates associated with PBF-L can produce deposits of iron-based metallic glass and may be a viable means to produce bulk material and parts (Pauly et al. 2013).

4.4.6 Recycled Metal

Recycling has grown to be a big business driven by a number of environmental and economic factors. Who can argue about the positive effects of recycling on the environment, sustainability, the impact on consumerism, and the ability of the world to accommodate a growing world population? We are all aware of recycling aluminum cans, but there is a lot of misperception about the effectiveness of recycling and reuse as it applies to AM. A good overview of metal recycling is provided in the Metals Handbook Desk Edition, pp. 31–3, (Boyer and Gall 1985).

The benefits of recycling are many. In one example, the remelting of recycled metal uses much less energy than primary metal extraction (mining and refining). Recycled metal can originate from two sources, "recovery metal" and "scrap metal", where foundries' remelt from these sources can be as high as 50% of primary commercial metal shapes. However, drawbacks and limitations exist. *Recovery metal* or *in-process scrap* is generated from well identified commercial metal stock resulting from conventional processing operations. Picture frames of cold rolled steel resulting from a metal stamping operation is one example, where sources of contamination resulting from the operation are well identified and minimal. *Scrap metal* or *post-consumer scrap* is harder to process as there are many sources of contamination that are difficult to remove during recycling, therefore limiting application for reuse. As one example, contamination of machining turnings by cutting lubricants limits the reuse of this type of scrap metal. Contamination from other sources, including rust, coatings, plastic, paints, galvanized metals (zinc or cadmium coatings) and lubricants will be incurred during the service life of the part. In addition, scrap metal has often lost its identity regarding the specific alloy of metal, limiting its reuse. There are methods to sort and classify scrap metals such as by color, weight, specific gravity, or modern test devices such as spectroscopy methods, but these can be expensive and only partially effective. As an example, the metal scrap industry has identified 20 categories of scrap steel and 10 for cast iron.[25] Identification and refinement of aluminum alloy scrap, such as machining swarf, often requires degreasing and removal of impurities using fluoride or bubbling chlorine through the melt. Therefore, the environmental impact of scrap metal recovery must be considered. Blending with new metal in the remelt is often required to achieve the desired metal requirements. In another example, aluminum beer cans can be made from two different alloys, one that forms easily into the can shape and another to provide strength for the lid.

4.4.7 Recycle and Reuse of AM Metal Powders

The high-performance engineering alloys most attractive to AM are not necessarily compatible for direct recycling into metal powder or wire for operations such as AM powder fusing. The pickup of contaminants in machining operations such as cutting fluids or contaminants picked up during in-service operations may disqualify the recycle of these expensive materials for direct reuse. For some reactive metals such as titanium, cost-effective recycle methods have yet to be developed. With a trend toward composite materials or components that utilize multiple metals, there may currently be no path forward for recycling. For these reasons the ability to recycle and reuse powder for AM processing is highly attractive on the basis of cost alone provided the purity and pedigree of these materials is preserved.

Impurity contamination of AM feed material can affect the chemistry and metallurgical response of the deposited material. Commercial grades of metal can feature wide ranges of impurity content associated with the degree of processing and are often reflected in higher costs for higher grade purity materials. In AM processing, build chamber atmosphere and delivery gases can contain impurities as

---

[25]Boyer and Gall (1985, pp. 31–34).

well potentially finding its way into the reused or recycled material. Storage of powders or wire and improper handling can also deposit impurities and contaminants that find their way into the build environment. Powder producers and OEM machine vendors are well aware of the importance of maintaining the purity and traceability of powder feedstocks and continue to evolve the processes and procedures needed to recycle and reuse these materials.

Reuse of metal powder produced for AM is one of the major attractions to AM technology, one that is heavily promoted by industry leaders, is that the technology is very ecofriendly using only the powdered material needed to make the part and allowing the reuse of all the remaining powder. As will be discussed in detail later in the book, most all AM metal process requires the use of some level of support structure which is later removed as scrap, some level of post process finishing such as machining and drilling with an associated waste stream and sifted process generated debris from the used powder before it can be reused.

Research is ongoing to improve and innovate ways to recycle industrial metal waste more effectively. LPW Technology in the UK has done studies to determine feasibility of recycling used powders. Another example of ongoing research shows the waste of one process can be salvaged and used directly as the input into the next (Mahmooda 2011). As mankind establishes a greater prescience in space, In-situ Resource Utilization (ISRU) will become an option to salvage and reuse materials launched from earth or utilize materials mined in space as the great cost to transport material out of the gravity well of earth is extremely costly. Conceptual planning by NASA and space technology developers for InSitu ReUse (ISRU) is already underway will add extra considerations for the design of components to consider material selection, in-service conditions and end of system life disassembly and reprocessing.

Now that you have a basic overview of metal structure, properties, metallurgy and form as they relate to AM, let us get some background on energy sources used by AM to melt or sinter these materials.

4.5 Key Take Away Points

- Metals alloys offer a wide range of useful properties such as strength and durability that are a result of the chemistry of the metal and the process used to form a part.
- The alloy chemistry and process used to form metal will determine the microstructure, properties and ultimately contribute to the performance of the part in service, under conditions such as high temperatures, corrosive environments or biocompatible applications.
- Metal alloy powders require special processing to achieve the desired spherical shape and purity needed by certain AM processes. This adds to their cost and limits the number of metal alloys commercially available for AM processing.

- Metal alloy wires and electrodes, widely used in welding applications, are being used for DED AM processes featuring bulk deposition.
- AM processing may also be used to form shapes using specialty alloys and composites, not easily formed by conventional methods.

Chapter 5
Lasers, Electron Beams, Plasma Arcs

Abstract High energy heat sources used to melt and fuse metal are widely used in industry. While laser and electron beam processing have been in use for 50 years or more, and arc welding for more than a century, the fundamentals of these processes are often poorly understood. The increased automation of these metal processing systems further remove the operator or engineer from the basic function and control of these heat sources, the molten pool and ultimately the fused metal deposit. AM metal processing systems often operate within the confines of an enclosed chamber and at high speeds creating extremely small molten pools further obscuring the basic functioning of the process. This chapter describes the basic function of these heat sources, what happens when a high energy beam or arc heats a metal surface to melt and fuse powder or wire into shaped parts. The use of auxiliary and additional heat sources during AM processing and post-processing is introduced.

High energy density heat sources under computer control can be put to work melting, sintering or fusing metal into functional 3D objects. This chapter covers the basics of melting, what heat sources are used in AM, why and when we use one or the other. Understanding the basics of these heat sources and how they are applied will assist you in selecting the process best suited to build the parts you need (Fig. 5.1).

5.1 The Molten Pool

When directing a concentrated energy source at a metal surface, some of the energy never reaches the surface and is lost, some is reflected as it hits the surface, and some energy is absorbed then radiated away as heat. The remaining absorbed energy heats the metal and if intense enough, creates a molten pool. The heat source may then be translated along a prescribed path, followed by the molten pool. Figure 5.2 shows the impingement of laser energy melting a substrate surface of metal. While the physics of the energy beam and molten metal pool interactions are

© Springer International Publishing AG 2017
J.O. Milewski, *Additive Manufacturing of Metals*, Springer Series in Materials Science 258, DOI 10.1007/978-3-319-58205-4_5

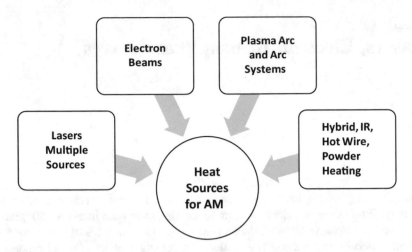

Fig. 5.1 Heat sources for AM metal

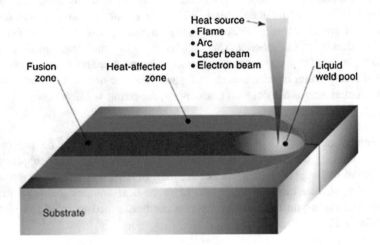

Fig. 5.2 Traveling heat source and liquid melt pool[1]

beyond the scope of this book, it is important to note that fluid flow within the molten metal pool and vaporized metal flow above the pool are very dynamic and difficult to control. As an advantage in AM, vigorous molten pool flow helps to break up surface oxide layers between powder particles and filler wires and assist the fusion into the previously deposited base layers. As a disadvantage, vaporized metal above the molten pool can be detrimental to the process, creating unpredictable process disturbances, loss of low vapor point alloying constituents and

---

[1]Figure courtesy of Lawrence Livermore National Laboratory, reproduced with permission.

condensation of vapors upon the optical elements or build chamber walls or filters, requiring removal by cleaning of replacement.

The beam and molten pool will melt and consume filler material, either in the form of preplaced powder, delivered powder, or wire introduced into the pool and fused to become the part. Fusion will progress by melting base metal and filler metal at the leading edge of the pool and allowing the metal to solidify at the trailing edge of the pool. In most cases where multiple deposition passes are needed, each pass must fully fuse into the previously deposited material.

Understanding the basics of melting, the heat source and the process being used will help you to design better AM components and correctly choose and apply the process parameters to attain the desired quality within the part. The reader is directed to this book's Appendix B where practical exercises using common welding equipment are provided to offer a deeper understanding and gain firsthand experience of how metal behaves when melted and fused. Reading about metal fusion is one thing, actually doing it and learning by experience is much more valuable.

5.2 Lasers

As mentioned earlier, the invention of the laser more than a half century ago was a technical milestone in physics that has found applications across a wide spectrum of material processing. Lasers powerful enough to melt metal were developed early on and found wide application in laser cutting as early as the 1970s, although equipment costs focused uses to high value, high payback production operations. Although the prices have dropped considerably, high-powered lasers used for AM systems can still cost hundreds of thousands of dollars.

Lasers generate a high energy density beam of *photons* and can be transmitted and focused to produce a small spot size of energy capable of melting and vaporizing metal. Laser equipment can generate beam powers of thousands of watts and focus to beam spot sizes of fractions of a millimeter. These small spot sizes can result in very small molten pools capable of melting at very high travel speeds on the order of many meters per second. There is a wide range of laser technology used in AM with names such as Nd:YAG, disk lasers, and direct diode lasers[2], but this

---

[2]High powered CO_2 lasers operating at kilowatt levels were the first to see wide use in metal welding and cutting. The 10.6 micron wave length required reflective optics such as copper mirrors and could not be transported in quartz fibers or windows, limiting their use during the development of the first laser based AM systems. Nd:YAG laser followed first with pulsed beams relying on pulsing flash lamps inside a reflective cavity to pump a laser rod of Nd:YAG doped quartz. Nd:YAG laser could be focused into and delivered by optical fibers and easily passed through windows into glove boxes, adding to their versatility. Flash lamps had a limited lifetime and required alignment and maintenance of optical elements. Lasers using multiple laser cavities and sequential pulsing to create quasi-continuous wave beams were also used in early AM systems. Optical diodes eventually replaced the flash lamps to pump the laser cavity, offering longer

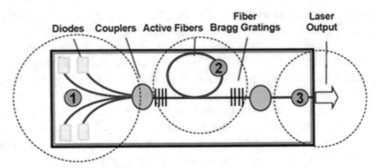

(1) **Pump diode modules** pump the light radiation into the active fiber

(2) **Optical active fiber** with a *doped core* (ytterbium) and couble cladding, where the pumped light excites the core

(3) **Transport optical fiber** bringing out the power from the module

Fig. 5.3 Fiber laser principal[4]

discussion will focus on the newer fiber laser technology because most AM systems now use fiber lasers due to their reliability, compact size, and low maintenance. An article by IPG Photonics Corporation provides a good overview of fiber laser principles and applications.[3] The basic principal of a fiber laser is shown in Fig. 5.3. Optical pump diodes are coupled onto an active laser fiber with a special reflective coating and Bragg gratings that reflect the laser light back-and-forth, along the length of the fiber, to create a coherent beam of light at the output of the laser. Beam delivery is often accomplished using additional optical fibers that provide a durable, flexible, fully enclosed beam path for delivery and containment of the light energy. These fibers are safety interlocked to shut down the system in the event of a breach of the delivery fiber. Final beam delivery includes optical elements and lenses to condition and focus the beam after it leaves the optical fiber. Manipulation of the beam is often accomplished by magnetically driven mirrors or CNC motion as discussed later in the book.

The laser beam is directed toward the work and focused at or near the part surface at power densities sufficient to achieve the desired degree of melting. Laser

---

(Footnote 2 continued)

lifetimes and reduced service requirements. Disk lasers offered benefits to beam quality and direct diode laser array offered high-efficiency lasers for metal processing such as cladding at the cost of beam quality. The advent of fiber laser technology offered significant improvements in robustness and cost in a compact size.

[3]Article from EDU.photonics.com, Fiber Lasers New Types and Features Expand Applications, Bill Shiner, IPG Photonics, http://www.photonics.com/EDU/Handbook.aspx?AID=25158, (accessed March 17, 2015).

[4]Courtesy of The FABRICATOR, an FMA publication; reproduced with permission.

energy impinging on the workpiece is either reflected away or absorbed into the part or filler material creating heating and melting. Laser absorption or reflection may be significantly different for different metals.

Without getting into the complexity of laser material interactions, it is important to note that different metals will absorb or reflect laser energy differently, such as titanium versus copper. Aluminum or silver have very low absorption coefficients while titanium has relatively higher absorption. The type or wavelength of laser energy affects the absorption as well. Yb:YAG fiber lasers will pass through quartz optical elements, such as the window of an inert processing chamber, a focusing lens or within a quartz fiber. In comparison, a CO_2 laser wavelength will couple with and melt quartz making it impossible to transmit using a fiber and difficult to deliver into an AM processing chamber.

In addition to the material and wavelength absorption dependence, the absorption of laser energy by molten metal is often much greater than for solid metal. Plumes of vaporized metal or plasma can form above a molten pool, absorbing laser energy and preventing it from reaching the molten pool. If that is not enough, the pressure of the rapidly expanding cloud of vaporized metal and superheated gas can create a depression in the molten pool trapping laser energy and further enhancing absorption. This depression often called a *keyhole* creates a vapor cavity that may extend deeply into the metal. This *keyhole mode* of melting (Fig. 5.4) can produce deep penetration but also may create defects such as porosity, spatter (also *balling*) or entrapped voids. In some cases and at certain laser wavelengths (such as with the CO_2 laser wavelength of 10.2 nm) partially ionized gases or plasma may form creating additional process instabilities. These transitions in absorption and melting between heating, conduction melting, keyhole melting and plume formation can be very abrupt and can result from very small changes in laser power, travel speed,

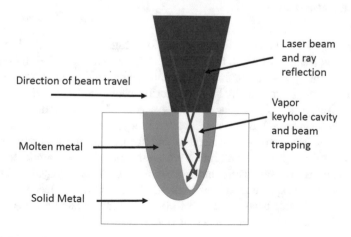

Fig. 5.4 Laser keyhole vapor cavity

focal spot size changes, or other minor process disturbances. In AM processing, powder size, and layer thickness can also have an effect. So how does this all work in AM? Careful selection and control of the wider range of laser, process parameters and materials as discussed later in the book are required to maintain a stable repeatable process.

As mentioned earlier, laser absorption increases upon melting and may be of sufficient energy to preferentially vaporize low melting point constituents of an alloy, such as aluminum, magnesium or lithium, thus changing the chemistry of the deposit. Vaporized metal may redeposit upon surfaces within the build chamber or nearby optical components. The dynamics of laser material interactions are extremely complex (solid, liquid, gas and sometimes plasma) and may be sources for process instability. Laser hazards, covered in the Appendix A and the ANSI standard publication (ANSI 2000) are typically controlled within the confines of the build chamber or glove box.

Books and standards related to development and application of laser welding technology provide good sources to understand the complexities of laser processing of metals and to follow the path taken to application and certification for the production of critical components. The good news is these applications have developed a large body of knowledge, compiled over four decades, to assist in the understanding and control of these processes. Recent developments in industrial laser systems have resulted in compact lasers that require low maintenance and can run reliably for tens of thousands of hours. These reliable, low-cost lasers have had a strong positive effect on the development and adoption of AM systems. Good reference books on laser welding technology are contained in these references (Ready 2001; Steen 2010; Duley 1999).

5.3 Electron Beams

As with lasers, electron beams (EB) also provide a high energy density beam but, in this case using *electrons* instead of photons. They can produce small focused beam spot sizes resulting in small molten pools and are capable of melting at very high travel speeds. The basic principal of an electron beam gun is shown in Fig. 5.5.

A high voltage supply is placed across a *grid cup* and anode. A negatively charged cathode is heated to boil off electrons in a process referred to as *thermionic emission*. Those electrons are accelerated and focused by the grid cup toward the anode passing through a hole and into a work chamber. In the chamber the charged electron beam is focused using electromagnetic coils and may be directed to locations on the workpiece using magnetic deflection coils to steer the beam. EB equipment can generate beam voltages of 60–150 kV and beam powers of 3–30 kV or more and focus to beam spot sizes of fractions of a millimeter.

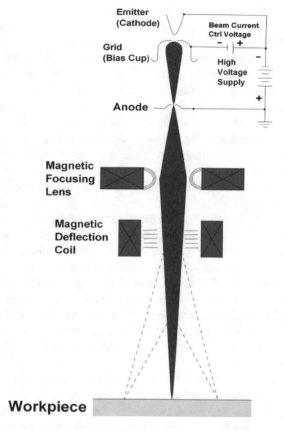

Fig. 5.5 Electron beam gun principal. "Schematic showing basic components and operation of electron beam materials processing," https://en.wikipedia.org/wiki/File:Schematic_showing_basic_components_and_operation_of_electron_beam_materials_processing.png#filelinks[5]

These all happens in a vacuum chamber providing a high purity environment (<1 × 10^{-4} mbar) within which to melt and fuse metal with minimal contamination by the oxygen and hydrogen present in air or moisture. From an AM processing point of view, the magnetic deflection allows very rapid traversal of the beam along a prescribed path. This beam deflection can be much faster than the magnetic scanning of mirrors used to position laser beams. A limitation of this process includes the limited life of the cathode emitter, requiring change out at intervals that may affect the maintenance and throughput of the AM processing of large pieces requiring significant beam powers and *beam on time*.

Industrial use of EB heat sources includes welding for aerospace, nuclear and high production applications that can justify the cost and complexity of these processes. Equipment costs have historically put these processes out of reach for entry and mid-level applications. EB systems rely on sophisticated computer control and computerized numerical control (CNC) motion systems. Equipment costs can range from hundreds of thousands of dollars to millions of dollars.

There are two types of electron generators referred to as low voltage (~ 60 kV) or high voltage (~ 100–150 kV) guns. Typically, the high voltage guns are fixed upon the vacuum chamber while the low voltage guns may be fixed to a smaller chamber or moved by CNC motion within a large chamber. Magnetic coils within the *electron gun* are used to focus and manipulate the beam within the chamber. More detail will be provided later in the book regarding specific process configurations as they relate to AM.

An advantage of electron beams is the ability to couple the beam energy more efficiently to metal than when using laser beams, due to laser beam reflection and energy loss. A disadvantage is that the electron beam needs to be generated and propagated within a vacuum as even small amounts of gas molecules will be impinged upon and diffuse the beam. Some electron guns utilize additional vacuum pumping directly on the gun to avoid contamination and back streaming of metal vapors. Some electron beams may be focused to beam spot sizes as small as, or smaller than laser systems typically used for AM. Electron beams impinging on metals can generate X-ray hazards that are typically controlled within the confines and shielding of the chamber.

As with lasers, electron beams may be of sufficient energy to preferentially vaporize low melting point constituents of an alloy, changing the chemistry of the deposit. As with lasers, vaporized metal may redeposit upon surfaces within the build chamber or electron gun components requiring regular maintenance. The dynamics of EB–material interactions are extremely complex (solid, liquid, gas, and plasma) and may be sources for process instability.

The development of industrial electron beam welding technology preceded that of laser welding. Books and standards related to development and application of electron beam welding technology provide good sources to understand the complexities of EB processing of metals and to follow the path taken from application through certification for the production of critical components such as aerospace parts (AWS C7 2013; O'Brien 2007; Lienert 2011).

Research continues to further evolve EB technology for AM processing. One such example is a project funded by the European Commission for the development of an electron gun with the ability to deflect and raster the beam more rapidly and at high powers to increase the speed of AM part processing.[6]

---

[6]The European Commission report Final Report Summary—FASTEBM (High Productivity Electron Beam Melting Additive Manufacturing Development for the Part Production Systems Market, Project Reference 286695, http://cordis.europa.eu/result/rcn/153806_en.html, (accessed March 17, 2015).

5.4   Electric and Plasma Arcs

An electric arc provides a high energy density heat source that can easily melt metal, although at a slower rate than laser or electron beam. Arc heat sources are fundamentally different from lasers but related to electron beams as a stream of electrons is transferred across the arc but at lower energy density than with EB. Whether you are using lasers, electron beams or arcs, the end goal is the same: to create a pool of molten metal. The arc characteristics can differ significantly depending on the direction of current flow (AC, DC+, DC−), or with additions to the arc environment, such as by choice of shield gases (e.g., nitrogen, argon) or the selection of filler metal when transferred across the arc. The physics of the electric arc and larger molten pool created by arc melting is described in (Lancaster 1986). A big advantage of arc heat sources is the lower equipment costs (in the range of thousands or tens of thousands verses millions of dollars) and greater wall plug efficiencies when compared to lasers or electron beams. A big disadvantage is the arc or plasma beam cannot be focused tightly (to below a few millimeters) to allow the accuracy and fine detail required by many AM applications. Deposition rates using the heat sources can be much higher and they can be used to deposit a wide range of commercially available filler materials at the cost of increased heat induced distortion and decreased deposition accuracy. Arc welding has evolved over the past century with a number of good references in print (Cutting 1980; Lienert 2011; O'Brien 2007) or information provided by the links to Welding Equipment and Consumables Suppliers section at the end of this book.

Plasma arc welding (PAW) is a variation of gas tungsten arc (GTA) welding in which a jet of inert gas is used to direct the welding arc into a focused high energy density arc plasma jet, as shown in Fig. 5.6. This jet of partially ionized plasma gas is able to enhance weld pool penetration by creating and extending a keyhole of metal vapor to the depth of the weld pool, enhancing penetration to a depth not attainable by simple heat conduction. The stabilized and directed arc of the PAW process offers a benefit when using complex CNC motion by constricting and focusing the arc, helping to prevent "arc wander." Filler material is introduced directly into the molten pool and is not an integral part of the heat source.

Gas metal arc welding (GMAW), also known as MIG welding, utilizes a continuous feed of filler wire as a consumable electrode. Metal is transferred across the arc into the resulting molten pool as shown in Fig. 5.7. Multiple weld passes may be used to build up layers of material or into shapes while under CNC motion control. Variations of the process include reciprocating wire feed (RWF-GMAW) (Kapustka 2015) and cold metal transfer (CMT-GMAW) (Furukawa 2006) helping to control heat input, reduce spatter and greatly improve weld beam shape control.

Robotic manipulation of the GMA weld process and computer control makes these processes contenders for the rapid buildup of large near net shape components by controlling the many variables such as current, voltage, polarity, arc length, torch angle, and travel speed. A good way to understand the fundamentals of melting at a scale you can see and learn to control is to pick up a GTA weld torch

Plasma Arc Welding (PAW)

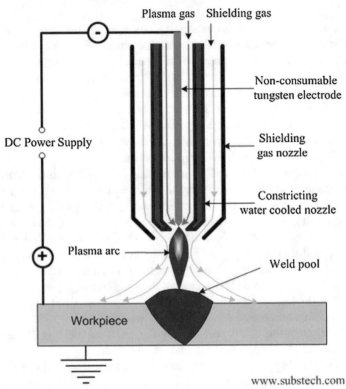

Fig. 5.6 Plasma arc welding principal[7]

and make some welds. Appendix B provides some examples of where to start with some hands on exercises.

Energy coupling and melting by arc heat sources produce molten pools much larger than those obtainable using lasers or electron beams. Large molten pool volumes are more difficult to manipulate and control when compared to laser or EB. *Surface tension* can create large rounded deposits or *weld beads* resulting in decreased deposition accuracy. However, deposition accuracy is less of an issue if the part is to be 100% post process machined. Gravity forces can sag or distort deposits while magnetic forces can deflect the arc position. The chemistry of the weld metal, inert gas cover, or arc environment may also affect the weld bead deposit shape and penetration. Having said all that, the low cost of arc welding equipment, the wide range of well characterized, certified materials and processes

---

[7]*Source* Dr. Dmitri Kopeliovich, "Plasma Arc Welding (PAW)," Subs Tech substances & technologies, http://www.substech.com/dokuwiki/doku.php?id=plasma_arc_welding_paw. Reproduced with permission.

Metal inert gas welding

(MIG, GMAW)

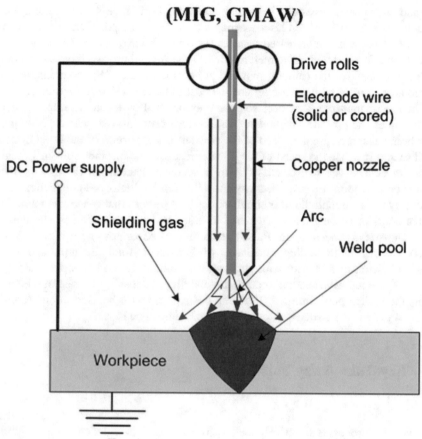

Drive rolls

Electrode wire
(solid or cored)

DC Power supply

Copper nozzle

Shielding gas

Arc

Weld pool

Workpiece

www.substech.com

Fig. 5.7 Gas metal arc welding principal[8]

and recent developments in process control make arc-based AM a strong contender
for certain applications such as rib on plate bulkheads, forging blanks, or hemi-
spheres for storage vessels or complex container geometry where *hog-out*
machining or expensive punches or forming dies is a less cost-effective option.

---

[8]*Source* Dr. Dmitri Kopeliovich, "Metal Inert Gas Welding (MIG, GMAW)," SubsTech substances & technologies, http://www.substech.com/dokuwiki/doku.php?id=metal_inert_gas_
welding_mig_gmaw. Reproduced with permission.

5.5   Hybrid Heat Sources

Arc and laser heat sources can be combined as a hybrid process and combine the benefits of each. Arc-assisted laser welding can combine the penetration and travel speed of a laser with the increased material deposition rates of an arc-based system. Laser beams focused within the melt region of a GTA weld process have shown the potential to stabilize the cathode spot and arc characteristics. This combination of hybrid heat sources may further reduce the cost of future AM metal systems.

The use of multiple laser heat sources may be used as means to improve deposition rates and provide a redundant heat source for certain AM systems. Multiple laser beams can increase the speed of the powder fusion portion of the build cycle, but it cannot increase the speed of the powder recoat cycle or other built-in system pauses or delays. We will discuss this in more detail later when discussing AM process optimization. Electron beam systems that feature high-speed deflection can effectively produce multiple near simultaneous heat sources that can create multiple molten pools or preheat the powder bed to assist in the distribution of shrinkage forces and stress concentration. Pulsed or modulated beam power for laser, EB or arc sources can retain melting efficiencies while reducing total heat input and may have application to AM. Hot wire applications, commonly used in weld cladding processes, use resistive heating to preheat metal filler, increasing the rate of melting within the molten pool. Infrared (IR) lamp heating may also be used in preheating the build platen or powder bed, enabling faster build speeds.

5.6   Key Take Away Points

- AM metal processes selectively melt feedstock materials under computer control into useful shapes based upon a computer generated solid model. This most often requires a directed energy source to melt or sinter the metal powder or wire, along a prescribed path, to fuse and form the object.
- A fundamental understanding the various heat sources used by AM to achieve melting is needed not only to appreciate the complexity of these processes, but also to properly select the right AM process for the specific material and part design.
- High precision heat sources such as laser or electron beams may be used for application such as jewelry or components with small or intricate features. Arc and plasma arc heat sources are being used for the bulk deposition of large near

net shaped parts that deposit large volumes of material and rely on 100% post process machining.

- Multiple high energy beam heat sources or hybrid heat sources have been developed to increase the build rate, enabling the production of larger components.

Chapter 6
Computers, Solid Models, and Robots

Abstract High powered computing hardware, software, and sophisticated computer-based sensing and control has migrated from the factory and entered the cloud and the home. We have become comfortable with computers, software applications, and computer-driven machines as they now exist in our homes in many forms ranging from smart phones to remote controlled toys. We now have access to low cost 3D printers in the studio, or down the street, and access to larger commercial machines over the Internet. It is useful to understand what the building blocks of these machines are and how they fit together to best choose AM metal printing capability. In this chapter, we discuss the range and capabilities of 3D solid model software, 3D scanners, and computer-aided design (CAD) that support AM. In addition, we introduce and discuss computer-aided engineering (CAE), computer-aided manufacturing (CAM), computer numerical control (CNC), and motion systems. Most importantly, we focus on the aspects of these technologies as applied to AM metals. You will be introduced to the STL or Stereolithography file format, originally developed for rapid prototyping of polymer materials and plastics, offering a simple solution to computer-defined surface geometry. The evolving need for interoperability and cross platform independence is discussed as well as the development of the 3MF file format.

3D computer-aided design (CAD) software has been in wide use for many decades but has seen wider adoption in the past 25 years with the increasing power and availability of computer graphics, solid model software and work stations made available to design engineers. More recently, the power and speed of computing hardware and software has extended the reach of this technology to personal computers and laptop systems. Computer-aided engineering (CAE) software for thermal, mechanical, and fluid flow analysis continues to evolve with ever more powerful algorithms, hardware and software advances. Computer-aided manufacturing (CAM) has evolved significantly over the past two decades to include multifunctional machines (e.g., multi-turret lathes), to hybrid work centers (e.g., machining plus inspection), to multi-axis computerized numerical control (CNC) of lasers (e.g., 5-axis laser cutting). While high-end engineering software remains

© Springer International Publishing AG 2017
J.O. Milewski, *Additive Manufacturing of Metals*, Springer Series
in Materials Science 258, DOI 10.1007/978-3-319-58205-4_6

costly and complex, mid-range and lower costs solutions are becoming increasingly available to small businesses, students, and makers allowing entry into this virtual world of 3D objects.

6.1   Computer-Aided Design

Computer-aided design software has evolved significantly over the past 25 years from a 2D computer-aided drafting tool to full 3D model generators and beyond to include model-based engineering tools capable of virtual assembly and 3D functional simulation. Sophisticated graphical user interfaces are provided to allow rapid definition of geometry and high-quality rendering. Creation of solid geometry using direct programing methods and using the principals of constructive solid geometry (CSG) are also available as an alternative means of creating solid models.

High-end professional packages, such as CATIA, ProEngineer, and Unigraphics tie into a wide range of manufacturing resource planning, enterprise management and documentation software as well as databases and records associated with inspection, quality, and testing. Products are being developed to document all aspects of a functional component from cradle to grave, including model design, product and process definition, inspection and in-service performance. Terms such as "part fingerprint" or "product DNA" are being coined to describe the future tracking of manufactured parts to assist in product tracking, improvement, evolution and to prevent copying, duplication, or counterfeiting.

Mid-range packages, such as those offered by Solidworks and Autodesk, provide a high degree of functionality at a reasonable cost. A wide range of commercial software, optimized for engineering, animation, architectural rendering or artistic pursuit, all provide the capability to input and output solid models in formats that may also be used for 3D printing. Mid-cost packages can offer precision drawing and modeling using a wide range of file types allowing imports and exports and allowing file sharing between software platforms. An important distinction is those models that offer *parametric* design. Parametric models are defined using variable values for dimensions rather than constant values. Modifications to these dimensions are easily made and reflected in modifications to the model geometry, without the need to recreate a new model. Feature-based solid models define geometric shapes that may be combined in ways to create more complex geometry. Parent/child relationships may be defined between these features or they may be defined as additive (e.g., a wall feature) or subtractive (e.g., a hole feature).

Low-priced software for both 2D drawing and 3D solid models can be obtained under open-source license for little or no cost. Free or low cost CAD software includes Google SketchUp, 3Dtin, Meshmixer, Autodesk 123, and Tinkercad. These packages enable beginners to create models easily, which can be converted to STL file formats and printed. Others supply entry level packages that are more capable products with greater functionality but also require a bit steeper learning

Fig. 6.1 A 3D solid model
surface represented as a series
of *triangles*[2]

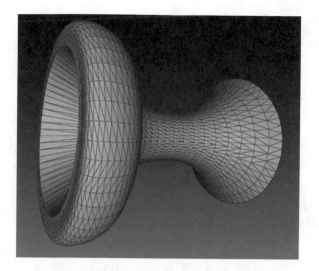

curve. OpenSCAD[1] is an open-source direct programming environment for creating
solid geometry that may be output as STL files for 3D printing. Appendix C provides a programming example that can be output as an STL file for slicing and
3D printing.

While many of these software packages have their own proprietary file formats,
they often provide translators into other formats used by other software, although
some information within these file formats may not be translated and carried along
providing only a limited level of translation. Limitations on translation are especially problematic when translating to the STL format most commonly used for
3D printing.

The STL file format (from StereoLithography) was created by 3D Systems for
rapid prototyping and stereo lithography. It represents the surface of a solid model
as a series of interconnected triangular flat facet surfaces as shown in Fig. 6.1.
Figure 6.2 shows the difference between a 2D CAD model representation of a
donut shape and a representation approximated by triangles to illustrate how and
STL file can approximate a surface. The definition of a single STL triangle in
ASCII text format is shown in Table 6.1. Very large ASCII text files may be
represented in a more compact binary format file. The greater number of triangles
used to approximate the surface the better the fit, but the larger the file size. The
triangle vertices are represented as three x, y, z, points in *Cartesian space* with a
vector pointing outward in relation to the surface of the geometry being described.
No edge may be shared by more than two facets and no two facets may occupy the
same space. All normal vectors must point outward and the represented surface of
all connected triangles must be fully connected or "water tight". This simplified

---

[1]OpenSCAD Foundation, http://www.openscad.org/ (accessed April 10, 2015).

[2]Unattributed image under CC BY-SA 3.0: https://creativecommons.org/licenses/by-sa/3.0/.

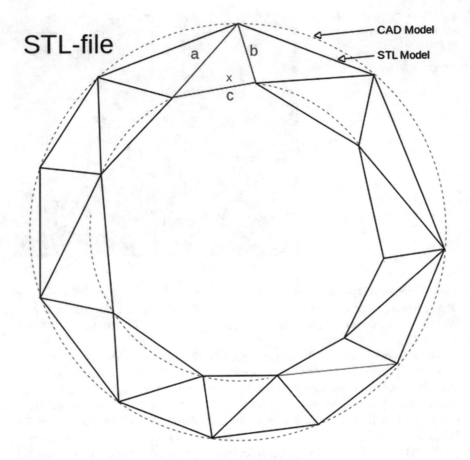

Fig. 6.2 The differences between CAD and STL Models, representing how STL modeling works. "The differences between CAD and STL Models," https://commons.wikimedia.org/wiki/File:STL-file.jpg[3]

surface representation carries no other information along with it, therefore additional information present within a CAD model will be lost when being translated to STL format, such as origin points, lines, datum planes, material type, color, dimensioning, or parametric feature relationships. Due to its simplicity when compared to solid CAD models the STL surface format has seen wide adoption. However, entry level software and computing platforms may limit the number of triangular elements that can easily be rendered and processed, ultimately limiting its use. The characteristic triangular faceted surface finish of some 3D printed consumer generated designs often reflected the limited resolution of certain STL

---

[3]Courtesy of Laurens van Lieshout under CC BY-SA 3.0: https://creativecommons.org/licenses/by-sa/3.0/.

Table 6.1 .

```
facet normal -0.998175 0.0531704 -0.0286162
  outer loop
    vertex -21.139 2.29811 -3.27318
    vertex -21.2163 2.00825 -1.11598
    vertex -21.2078 2.17906 -1.09682
  endloop
endfacet
```

surface models. An additional limitation of the STL format includes problems with scaling to high resolution and or lattice forms. STL files also cannot be translated back into the native CAD files of their origin. A 3D printer control file converted from the slicing of an STL file example is given in Appendix D. Figure 6.3 compares two sequences in building a part from a STL surface model or a solid model. Each sequence may start with a 3D design or scan of a 3D object. Supports are added to the STL defined surface model, it is sliced and used to build a part layer by layer. The solid model is used to define a CNC tool path CAM model from which the part is built. A CNC toolpath may define either a layer-by-layer deposition path or a feature-by-feature path utilizing 3D simultaneous multi-axis movement. Greater detail will be provided later in the book.

The *Additive Manufacturing File* (AMF) format (Lipson 2014) is a new standard being developed to enhance the attributes of an STL format by adding colors, multiple materials, curved surface representation, and other CAD features. AMF also features backward compatibility with the STL format along with improvements in performance, file size, read/write time, accuracy, and extensibility. Metadata fields may be used to include other information such as identification text.

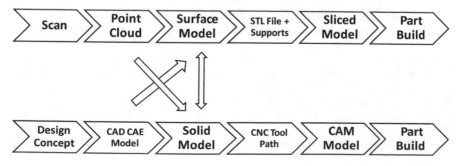

Fig. 6.3 Process flow for model creation and building a part

Parametric relationships within native CADfiles are not preserved nor are translation back to native CAD surfaces such as those defined by Non-uniform rational B-spline (NURBS) surfaces. The NURBS mathematical model is commonly used industry wide to precisely describe surfaces and curves in CAD, CAE and CAM software applications and is used within CAD file standards such as IGES (Initial Graphics Exchange Specification) and STEP (Standard for the Exchange of Product model data). The new AMF standards are being proposed and developed by national and international standards organizations such as ISO, ASME, ANSI, and ASTM F42 Committee.[4] The European specification ISO TC261[5] will include provisions for attaining more accurate surface representations as well as material differences such as color and ISO/ASTM 52900 Additive Manufacturing—General Principals and Terminology.[6] Understandably, there has been some resistance by AM machine vendors to yield to new standards, open source their proprietary software to allow compatibility or incur the cost of frequent revisions.

AM software specific to the control of metal powder bed systems may have additional requirements than those used for plastics. But since software development can be as costly and time consuming as hardware development, it is not surprising that software development for AM metal processing systems is proceeding cautiously. Given the dynamic nature of the AM market, small agile developers will have an advantage over large commercial CAD/CAM vendors that move slowly to incorporate process specific features into their base products. Third-party add-ons, linked to the most widely used commercial packages, may provide an interim solution but the interfaces between large commercial CAD/CAM software will always lag the newest capabilities of AM.

In addition, new file format designed for use with additive manufacturing, named 3MF, has emerged from a consortium of large CAD software, AM hardware, and AM service providers to address the limitations of the STL and AMF file format. It offer increased functionality and streamlines the process of design to 3D printing without interim file translation while preserving the information fidelity within the model. It is designed for use across a wide range of software and printing platforms.[7] The 3MF Consortium has seen a wide increase in membership since being announced in 2015. Open-source code development and cross platform interoperability is rapidly evolving.

3D Scanning hardware uses laser-based sensing or multiple images to create a *point cloud* of data approximating the spatial extent of an object within a defined coordinate system. Computer algorithms are used to smooth the data and render an

---

[4]ASTM International Committee F42 on Additive Manufacturing Technologies, http://www.astm.org/COMMITTEE/F42.htm (accessed March 19, 2015).

[5]International Organization for Standardization, ISO TC261 Additive Manufacturing, http://www.iso.org/iso/iso_catalogue/catalogue_tc/catalogue_tc_browse.htm?commid=629086 (accessed March 19, 2015).

[6]ISO/ASTM 52900, Additive Manufacturing—General Principals and Terminology, https://www.iso.org/obp/ui/#iso:std:iso-astm:52900:ed-1:v1:en (accessed April 11, 2016).

[7]3MF Consortium Web link, http://3mf.io/ (accessed January 28, 2016).

approximate surface representation which may be translated into CAD surface or solid model format. An article by: (Chang 2013) provides a detailed overview of the process needed to go from a scanned point cloud to a parametric solid model capable of translation and export to leading CAD packages such as CATIA or SolidWorks. 3D scanning software is often capable of taking a point cloud and directly fitting a surface with triangles to define the shell of a solid, creating a file output in STL format subsequently allowing the use of other software for STL model slicing and 3D printing. In other cases, manual interaction and editing may be needed to fully enclose and define a surface model or to define and fit parametric geometric solids, representing and combining features of a 3D model.

3D scanning and the associated software can be used for the creation of original 3D models based upon objects and reverse engineering applications based upon existing parts. A Geomagic Community example[8] demonstrates the ability to scan an automotive gearbox case and process the point cloud data into a model that can be cleaned up, closed up, features defined and then use to create a 3D model for printing a plastic or metal component.

Parametric model design features are based on solid shapes that may be combined by Boolean operators into more complex shapes to capture design engineering intent and fabrication details. Such features may include datum reference planes, base features (e.g. bodies or cases), additive features (such as bosses, ribs, flanges) or subtractive features (e.g., hole, slot, or machined surface) to be added or cleaned up by editing to achieve final dimensions during finishing operations.

Parametric design allows defining the relationship of all the features to one another in a tree structure or hierarchy of parent and child features. This can allow rapid design changes, scaling, dimensioning, the addition of more features, feature suppression, and the creation of feature-based additive or subtractive tool paths. High end professional CAD software allow the creation of technical data packages that travel with the solid model providing more options to preserve information and increase to utility of the design.

6.2 Computer-Aided Engineering

Computer-aided engineering (CAE) refers to the process of performing engineering analysis using sophisticated computing software, to aid in process simulation, prediction, and optimization. Analysis of thermal, mechanical, or fluid dynamic effects are in wide use as enabled by these tools. Analysis of heat flow and of deformation mechanical systems, such as the development of residual stress due to thermal or mechanical forces, are of particular interest to the development of AM

---

[8]Geomagic Community, Rebuilding a classic car with 3D scanning and reverse engineering, http://www.geomagic.com/en/community/case-studies/rebuilding-a-classic-car-with-3d-scanning-and-reverse-engineerin/ (accessed March 19, 2015).

Fig. 6.4 Finite element analysis of a moving heat source[9]

metal fabrication processes and prediction of part performance. Scientific computational models (mathematical models) of the behavior of metals exist at multiple spatial scales ranging from the interaction of atoms and molecules to the performance of large scale components and systems. The development of metallurgical process models such as the evolution of microstructure during solidification, grain growth, distortion, or stress are all active areas of research at corporate and national laboratories and universities. Simulations use one or more computational models to predict the behavior of a system. Commercially available software provides a useful interface to allow engineering analysis of computer-based geometric models such as those used to define CAD models using simulations based on computational models. Figure 6.4 shows a simulation of the heat flow in metal surrounding a moving heat source and melt pool during welding.

Model-based engineering (Fig. 6.5) can enable a wide range of engineering analysis to be performed to understand thermal, mechanical, fluid flow condition and make changes to the design to assist in optimizing the performance of the part. Finite Element Analysis (FEA) has been successfully applied to melt processes such as casting and in some cases welding. FEA takes a CAD model of a part or a sample and divides it into small volume elements or voxels. It then applies boundary conditions, such as the thermal and mechanical conditions, such as those present in fabrication or in-service conditions and performs calculations to simulate

---

[9]Courtesy of the University of Alberta, reproduced with permission.

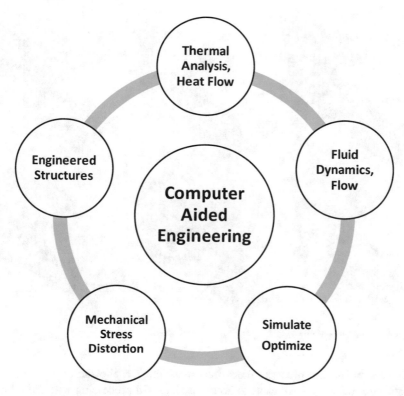

Fig. 6.5 Computer-aided model-based engineering analysis

the behavior of the system over time. Model-based engineering offers insight into performance of designs and can in some cases effectively serve as a tool to speed the design to prototyping cycle.

FEA of additive processing using lasers or electron beams presents a greater challenge as the scale of mesh refinement needed to accurately model a moving heat source and powder interaction adds to the computational challenges and computing power needed to obtain a solution. Other AM simulation challenges include the very small size of the pool and the high speeds associated with the build process. Material property data at temperatures ranging from ambient to vaporization is limited, thus limiting the accuracy of first principal models and requiring the assumption of missing data. The interaction of the many AM parameters and the effect of complex shapes, thin and thick walls, support structure restraint and metal to powder boundary conditions are highly complex and in some cases poorly understood. If these FEA engineering tools take longer to run than to rapidly prototype and test a part or if optimization results in a design that cannot be built because of AM process limitations then the value of the software and the time it

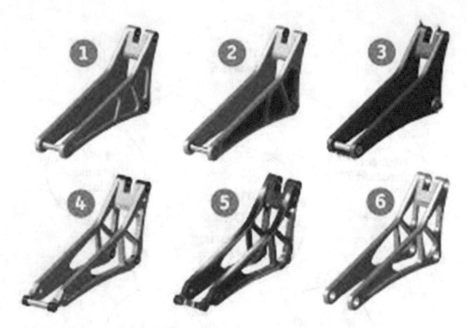

Fig. 6.6 Topology optimization[12]

takes to simulate the process comes into question. Despite these challenges, the benefits of modifying CAE tools to accommodate AM processing will undoubtedly drive the continued evolution and application of FEA technology.

One such application of CAE being used for AM is referred to as topology optimization. It is a CAE process by which mechanical loads may be applied to a design to determine load paths and guide the removal of excess material from the design while retaining the design strength. Iterative optimization through simulation may then be applied, resulting in strong lightweight designs, as shown in Fig. 6.6. AM could benefit from topology optimization methods to create complex designs used to fabricate complex geometry (Brackett 2011). Software featuring topology optimization, such as *Altair Hyperworks OptiStruct*, has become commercially available and can perform load path analysis and automatically perform and evaluate multiple design simulations to optimize weight and sizing of metal and composite structures.[10] Another optimization software is *solidThinking Inspire* has a web site[11] with an application example and videos using the software to create

---

[10]Altair Hyperworks, OptiStruct software, website, http://www.altairhyperworks.com/ (accessed March 19, 2015).

[11]SolidThinking Web site, http://web.solidthinking.com/additive_manufacturing_design (accessed March 19, 2015).

[12]Image courtesy of SolidThinking Inspire, reproduced with permission.

lightweight components. In one example, HardMarque has created a redesign of lightweight titanium pistons and in another Renishaw with Empire Cycles features the design of a lightweight AM printed bike frame. Computer-optimized designs such as these may not take into account critical flaw size or detrimental surface finish conditions where small thin lightweight ligaments may be at a greater risk of failure due to flaws and defects, resulting in a less robust design. Future case studies of the testing of these components will certainly help define and improve this design space. This capability takes us out of the box of conventional design thinking and into a new field of CAE design. Evolution of these software capabilities will surely follow the evolution of AM driven by the benefits to industry.

6.3   Computer-Aided Manufacturing

Computer-aided manufacturing (CAM) is the process of taking a CAD file, using the part definition to create instructions for a CNC machine tool, to perform the motion and machine control instructions to produce a component. As described above, CAD programs can produce solid or shell models describing the geometry of a part to be fabricated. The CAD file and associated information may be output in various file formats. These files need to be translated into instructions used to control specific machines such as 3D printers or CNC equipment such as lathes, motion equipment, or robots. The translation program is referred to as a postprocessor program and the translation process often referred to as "posting the file". Commercially available CAM software is offered in a wide range of prices associated with its capabilities. Some CAM software retains the parametric relationship of the CAD solid model allowing design parameter changes to be propagated into the CAM model enabling rapid regeneration of a new toolpath and machine instructions. High-end CAM software platforms can create toolpaths for a wide range of lathes, milling machines, laser cutters, and multi-function work cells. High-end systems for conventional processing also provide simulation of material removal, obstacle avoidance and tool shape simulation. CAM files can preserve the parametric relationship to the original CAD file allow for greater ease in modifying the original design, such as scaling or changing a dimension and re-posting the file to obtain a new toolpath.

CAM files contain machine control "M" codes and motion control or "go" codes, referred "G" codes. M and G code generation is defined within the postprocessor for each specific CNC machine. The STL example in Appendix D shows M and G codes for a simple AM plastic printer. This CAM programming and file generation process is more complex and less automated than creating the STL files commonly used for 3D printing, but the CAM files formats are more powerful as much more information can be carried in the file to the work cell. AM applications may require using multi-axis milling or machining sequences provided by the CAD/CAM software and mirror or reverse a tool path to enable use by an AM machine. Some manufacturers are beginning to add capability to their software to allow the definition of additive build sequences as well as conventional subtractive toolpaths.

As will be described in more detail later, directed energy deposition (DED) AM machines, using wire or powder feed, operate more like a standard CNC machine tool, but instead of removing material such as in lathes or milling machines they add material. They define a path to trace the volume of a part or feature to be deposited either on a layer-by-layer flat surface or upon a complex or contoured surface. The power of additive DED motion control when combined in hybrid systems with CNC subtractive tool control and in process inspection has been demonstrated and is now making its way into commercially supported systems.

CAM is also used to refer to the process of slicing STL files and translating them into 3D printing machine instructions used for the layer-by-layer buildup of the part. Powder bed type 3D metal printers use a CAM process similar to plastic 3D printers. They start with an STL file format, generated by their CAD software, orient the model into a defined build space, design support structures, slice the model into a series of planar layers and imbed scan parameters such as hatch spacing, step height, scan speed, contouring operations, heat source power, etc. Scan parameters associated with the 2D (X–Y) position of the beam within a specific slice are referred to as G codes. They functionally define the position of the beam and where to "go to" next. Other machine function codes, called M codes, set and change machine control parameters such as beam on, beam off, traverse speed, etc. The machine controller sequentially executes one line of commands after the next until the process is complete. Some manufacturers hold as proprietary their specific postprocessors and algorithms used to generate raster paths and commands specific to controlling their model of 3D printer. Various schemes to improve accuracy of contour paths (skin) and improve deposition rates as in fill paths (core) are employed, although the details are not always accessible to the user. These protections may inhibit the ability of the user to change or modify system parameters to suit user defined build requirements. Software may also utilize security measures, such as encryption, to protect proprietary information or enforce the use of proprietary branded materials. We will revisit these discussions later in the book when describing the process to design an AM scan sequence and control sequence.

Software specifically designed for 3D printing and additive manufacturing provides additional utility beyond that of CAD systems. They can read in various file formats, such as STL, enable scaling and provide the utility for slicing, hatching, or texturing. They can also support the generation of support structures

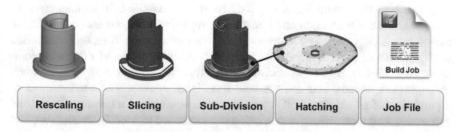

| Rescaling | Slicing | Sub-Division | Hatching | Job File |

Fig. 6.7 Bridging the gap between 3D software and printer[13]

needed to support the part while being 3D printed. One such example is shown in Fig. 6.7 where the Materialise Magic build processor module assists in the processes needed between design and a build job file and those specific to a 3D printer or AM machine. We will revisit this topic again later in the book.

6.4 Computerized Numerical Control

CNC machine tools and equipment are precision machines operating under the instructions generated by CAM software or by direct programming using M and G codes. This equipment may be conventional subtractive processing systems, such as lathes, mills, and inspection systems or additive systems such as 3D metal printing machines using powder or wire with arc or laser heat sources. As mentioned earlier in the book, CNC machines have been with us for over half a century. The earliest machine ran on programs stored and retrieved from punch tape but evolved along with computer controls and networking technology to be highly programmable precision systems. Systems can be as simple as a single rotational horizontal or vertical axis to those with over ten axis of motion interfaced with material handling, inspection and monitoring systems. Systems can be used for discrete part fabrication or function as continuous material processing lines. What sets CNC motion systems apart from other mechanical systems providing motion control is their ability for complex programmable control, high precision, absolute positioning and the ability to execute large CNC command files. Absolute positioning refers to positional feedback confirming motion command execution. As an example, the highest precision systems can fabricate parts to accuracies and surface finishes orders of magnitude greater than the best AM deposited parts made by melting or sintering.

---

[13]Courtesy of Materialise Build Processor, reproduced with permission.

3D metal AM systems based on CNC control are available in various configurations. Some are built upon lathe or milling type platforms, others on CNC controlled precision gantry systems and others simply rely on computer controlled configurations of precision mechanical motion components interfacing with computer controlled laser, arc or electron beam systems. Hybrid AM systems often integrate AM directly into conventional machining centers. The enhancement or repurposing of an existing CNC system may now integrate a compact laser head and power feed mechanisms and is being demonstrated to modify, modernize, and enhance existing processing lines.

Low precision, low-cost AM motion systems typically operate under computer control but are not generally referred to as CNC systems. One example is an XYZ gantry-type system, another is a *Rep-Rap*-type articulation system. This type of motion can be integrated with a low-cost wire feed arc welding system and these open-source designs have been demonstrated as described in Appendix E. Complex articulated motion and advanced sensing as provided by robots is also being applied to AM systems.

6.5 Robotics

The term "robot" has its origin in the configuration of a machine to have human like attributes such as an "arm" or autonomous motion. In the context of AM and CNC, we refer to robots as having a high degree of articulation and flexibility combined with advanced sensing and control. This blurs the definition of what looks like a robot (arms, head, etc.) but is in line with current vernacular. Robotic arms used for assembly, welding, weld cladding, and weld buildup have advanced significantly in recent years and may now be coordinated with other robots or robotic functions to include material transport, handling, remote operations and semi-autonomous operation. In the case of 3D metal printing and shaped weld buildup, these advances in automation are what are required to fully accommodate the needs and demands of complex 3D design shapes, large objects, and complex deposition paths and moveable or remote processing platforms.

Robotics and remote control technologies have now matured and have been reliably demonstrated and used for mass production. While not offering the highest precision attainable from machine tools, they have a strong track record of use in harsh manufacturing environments. Figure 6.8 shows an arc welding robot in action.

Like 3D printers, robots have captured the popular imagination as tools that makers and hobbyists can purchase, modify, and customize to serve their own

Fig. 6.8 Industrial robot performing arc welding[14]

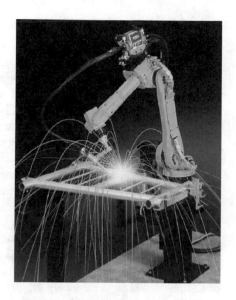

needs. As generalized automation platforms they can function as highly automated flexible machines for GMA welding large components. They are fully able in their current configuration to deposit complex-shaped weld buildup of very large parts, using a wide range of commercially available materials. Machining would be required to produce a finished part. Unlike AM/SM hybrid systems, the robots themselves are currently lacking in the precision, sensing, and controls of a CNC machine tool to perform precision machining operations, but research and development is being performed to achieve these goals.

The Delcam COMET PROJECT Webpage provides a good description of challenges preventing robotics for use in machining applications[15] and AM use.

From a conceptual point of view, industrial robot technology could provide an excellent base for machining being both flexible and cost efficient. However, industrial robots lack absolute positioning accuracy, are unable to reject disturbances in terms of process forces and lack reliable programming and simulation tools to ensure right first time machining, once production commences. These three critical limitations currently prevent the use of robots in typical machining applications

---

[14]"Arc-welding", https://commons.wikimedia.org/wiki/File:Arc-welding.jpg. © Orange Indus, FANUC Robotics Deutschland, CC BY-SA 3.0: https://creativecommons.org/licenses/by-sa/3.0/.
[15]From the Delcam and COMET PROJECT Webpages, http://www.vortexmachining.com/projects/comet.asp. http://www.comet-project.eu/results.asp#.VMZ5-i5i92E (accessed March 19, 2015).

6.6 Monitoring and Real-Time Control

Semi-automated motion systems have been around for the better part of a century. In some cases they evolved into sophisticated electromechanical systems for industrial production, automotive and aerospace applications. Some people in industry today recall the days when a CNC lathe was operated by punched tape in open loop control, operating as directed by control functions, with no control feedback aside from some safety limit switches that might trigger a shut down in the event of a problem.

Controller technology took a big leap forward with the advent of the micro-processor and programmable control offering a wide range of control options and the ability to incorporate sensor feedback into the control sequence.

Metal processing with arc, lasers or electron beams can create a harsh, highly dynamic, and poorly observable environment particularly when contained within the confines of an inert or vacuum build chamber. Extremes in temperatures, light, radiation, and electrical noise may restrict the observation of the process. Process conditions can change extremely rapidly in the very small, localized regions of the heat source and molten pool or within the build environment potentially disturbing or degrading system performance.

Monitoring of conditions such as part temperature, beam power, motion system function, or material feed conditions provides only a glimpse of the complex if not chaotic conditions during high energy density melt processing. The availability and adoption of sensors, multi-channel, high speed, computer-based data acquisition and analysis software help us understand the interactions of the fundamental pro-cessing parameters and better understand the process operating under normal conditions. Collecting this process data allows the development of feedback control systems and algorithms to react to process disturbances or abnormal conditions. Many of the lasers, EB, arc, motion system, and subsystem manufacturers provided monitoring and control interfaces allowing integration into an AM system. Monitoring and recording systems are often offered by the AM equipment manu-facturers as a system option.

Advances in high-resolution real-time digital camera monitoring provide a real-time view of the process and are able to filter noise, and to extract and record data while remotely displaying relevant images and data. These techniques have been hardened to withstand the harsh extremes of visible light, IR and UV radiation and protected from the build chamber environment. Image storage systems have evolved to allow both the display and capture very large datasets. Spatially resolved thermography provides an unprecedented view of thermal processing and may be used for process diagnostics, forensics, and model validation. Many of the AM machine vendors offer video monitoring options often adapted from other industrial monitoring operations such as welding. Some of the AM system manufacturers offer an open architecture allowing users to interface their own monitoring systems and use their own analysis software.

Laser profiling systems are used to determine the spatial intensity of the laser beam and can diagnose off-normal conditions related to the function of the laser or laser beam delivery optics. Similar systems are in use to characterize and diagnose electron beam systems as well. These systems, when applied correctly, can be used to assure proper beam delivery to the work chamber and detect degradation of the beam quality indicating either a required maintenance condition or an off-normal operation of the beam energy source. Pyrometer-based sensing may be used to measure temperatures at specific locations and times during the deposition cycle. AM machine builders and third-party vendors are teaming up to provide layer-by-layer monitoring of thermal conditions as the part is being built with the goal of assuring consistent and desirable deposition condition of each layer as the part is being deposited. Cameras and optical sensors may be installed within the laser delivery optics to sense melt region conditions to provide additional information regarding process performance. Laser scanning systems are being developed to provide rapid determination of surface uniformity or dimensional conditions for both powder bed fusion and directed energy deposition AM processes.

For arc based systems, noncontact process monitoring and control, such as through the arc sensing of arc length, may be used to control and steer the weld torch along the weld path. New power supply technologies have reduced the size and cost of the power supplies while providing rapid real-time control of arc current and voltage also providing real-time display and monitoring of primary process parameters. Variable polarity pulsed power supplies are another example of computer based controls that offer significant ability to tailor the arc conditions to the materials and parts being welded. High speed wireless multichannel data acquisition is sufficiently hardened to withstand the extremes of the AM processing environment for arc based systems.

6.7 Remote Autonomous Operations

Technology has evolved to reach truly remote locations with probes, robots, and automation relying on the fabrication of small and large metal structures. Landing on comets, 3D printing structures on a space station or using robots to repair a deep water drilling structure are all realities. Closer to home, pulsed gas metal arc welding utilizing weld current control is being demonstrated to perform 3D weld buildup and repair of complex shapes in remote locations such as on ships at sea serving oil rigs or within nuclear exclusion zones such as in nuclear reactor repair. The robotic arc deposition process limits the heat input to a part and controls the melt pool to the extent that freeform multi-axis deposition is made possible. The technology has demonstrated unsupported 3D free space deposition. In a look to the distant future, a robot, working alone in a remote location, may be able to help form a functional structure (Fig. 6.9) or object, or autonomously evaluate and perform a repair.

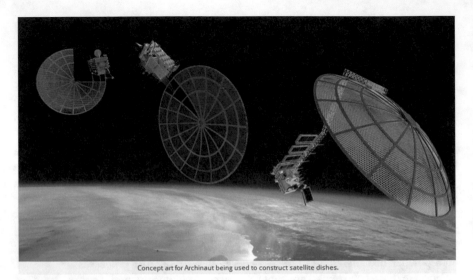

Fig. 6.9 Archinaut Progression[16]

Having said all that, a question to be asked is which monitoring sensors and data collection do I need? We will provide answers to these questions later in the book specific to AM processes and applications. By now you have a basic introduction to the software and hardware subsystems integrated into an AM system. In the next chapter, we will take a look at some of the precursor technology from which the knowledge base used to develop AM metal technology was derived.

6.8   Key Take Away Points

- Software used to generate computer based solid models in support of AM is readily available for CAD, CAM, CAE and CNC applications. The cost and capabilities of this software range from free open-source learning tools to large complex engineering and product lifetime management systems.
- New formats such as AMF or that being developed by the 3MF consortium are extending the functionality of the STL file format used by 3D printing and much of AM processing for the past 25 years.
- Additional third-party software, used to generate lattice structures, AM support structures and complex surface conditions are being offered as add-ons and are being integrated into to existing software packages in support of AM.

---

[16]Courtesy of Made In Space, reproduced with permission.

- Engineering software tools developed to analyze and simulate heat flow, fluid flow, mechanical performance, or optimize the topology and shape of lightweight designs are being applied to AM designs.
- Advanced process monitoring and in-process quality controls are being developed and integrated into production systems with a goal of real-time process control in support of quality assurance and process certification.

Chapter 7
Origins of 3D Metal Printing

Abstract 3D printing and additive manufacturing brings together and continues to draw from advances within a wide range of technologies such as information, computing, robotics, and materials. Developments in all of these highly visible, high impact, and highly publicized sectors will undoubtedly be modified, adopted, and integrated into the evolution of advanced manufacturing. Advancements within technologies smaller in scope, such as 3D printing plastics, or less visible sectors such as powder metallurgy, laser and weld cladding will also continue to play an important role in the continued evolution of AM metal. It is useful to understand the origins of AM metal processing as derived from these technologies as they will continue to play an important role.

AM has its origins in a number of base or precursor technologies (Fig. 7.1). Some of these have been with us for 20 years, some for half a century or more. Those widely applied in manufacturing are still evolving with new applications within and outside AM technology. We mention these important technologies, because large databases and experienced workforces, just outside the reach of the mainstream AM metal community, hold a wealth of knowledge relevant to AM metal processing. As AM metal development races ahead let us not forget someone may have already developed a solution to our problems using a different material or similar process. Therefore it is instructive to review a few of these technologies and consider their technical trajectory as it may apply to AM metal.

As introduced earlier in the book, lasers were invented over a half century ago. The acronym LASER (Light Amplification Stimulated Emission Radiation) has entered into the common vernacular. Lasers transform energy into a highly ordered (*coherent*) beam of light within a narrow wavelength that can be formed into a beam of photon energy, transmitted, directed or focused. Common industrial lasers are based on gas or solid state lasing mediums (such as CO_2 or doped crystal laser rods), pumped by an optical source to generate the beam. Modern lasers of the past decade based upon fiber laser and diode laser technology have seen significant improvements in laser power, cost reduction, reduced system complexity, smaller system size, increased robustness, and improved laser beam quality. Lasers have

© Springer International Publishing AG 2017
J.O. Milewski, *Additive Manufacturing of Metals*, Springer Series in Materials Science 258, DOI 10.1007/978-3-319-58205-4_7

Fig. 7.1 Origins of AM
metal processing technology

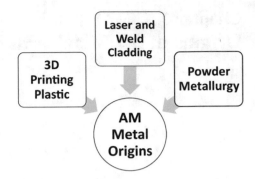

become cheaper, more powerful and easier to use. These laser benefits, along with increased integration with other advancing fabrication technology, such as CNC, advanced computerized control and sensors have allowed lasers to make significant inroads toward displacing historical fabrication methods in industrial prototyping and production environments. Laser cutting is one such example and as you will see, laser cladding is another.

Applications using laser drilling, glazing, and surface modification have all found application in industry. Laser machining and ablation have also been demonstrated in certain applications although the lasers used for these applications are significantly different from those used in AM metal processing. Despite all these improvements, high-powered lasers remain costly and for reasons of safety and security are not for everyone, but as you will see the benefits can outweigh the costs.

7.1   Plastic Prototyping and 3D Printing

Building and testing of prototypes has always been an important step in settling on a final design. In the past fabrication of functional prototypes was a slow and expensive process as it often required a number of iterations to achieve a functional part worthy of testing under actual service conditions. The advent of 3D printing with plastics and polymers created the rapid prototyping processes in wide use today and continue to serve as a precursor technology to 3D metal printing. Figure 7.2 shows the high-level process flow for 3D printing. Stereolithography (SLA), was invented by Charles W. Hall in 1984 and commercialized by 3D Systems in 1989. SLA uses UV light to cure photopolymer into 3D shapes and the process is often cited as the origin of 3D printing, as shown in Fig. 7.3. Selective Laser Sintering (SLS) was developed by Dr. Carl Deckard and Dr. Joseph Beaman at the University of Texas at Austin in the mid-1980s with commercialization by DTM and later acquired by 3D Systems. The technology uses a laser to fuse powder within a bed of material (plastics, metals or ceramics) into 3D shapes. Fused

Fig. 7.2 Process flow for 3D printing CAD to part

Deposition Modeling (FDM) was developed in the late 1980s by S. Scott Crump and commercialized in 1990 by Stratasys. FDM extrudes a thermoplastic through a heated nozzle to deposit planar layers into a 3D part, shown in Fig. 7.4. A comprehensive overview by the Science and Technology Policy Institute, describes the origin of 3D Printing and Additive Manufacturing, identifying and linking the top 100 AM patents and the evolution of the technology.[1]

Vanguard companies such as 3D Systems and Stratasys continue to pioneer methods to realize 3D shapes and develop new materials. These processes begin with a 3D CAD model saved as an STL 3D surface, which is then sliced and prepared for printing, then translated to machine instructions to control the buildup of multiple layers of materials to create a finished shape of plastic or paper. Initially these models were good for form and fit testing as well as marketing but have evolved to produce fully functional parts. Many variants of this technology pioneered the use of lasers to fuse or sinter layer upon layer of plastic powders and liquid polymers into parts. Binder coated metal powders could be fused into porous metal shapes then infiltrated with another lower melting point metal to form a solid part. Other prototyping technologies would laminate layers of material, such as paper, into 3D shapes. As will be mentioned later, the technology continues to evolve and find applications and markets.

Plastic prototyping today is primarily divided into two groups (1) a powder or liquid bed based system (refer to Fig. 7.3) fusing or curing the material using a laser or heat source and (2) those depositing material by extruding through nozzles or by print heads (refer to Fig. 7.4). Both start with a 3D computer model, slice it and build a part one planar slice at a time. Figure 7.5 lists some pros and cons of 3D printing plastics, polymers, and composites although improvements are continually being made particularly in size and material options.

As mentioned earlier, an enduring standard that evolved from this technology into the 3D printing of today was the rendering of 3D surfaces using tessellated (triangulated) surfaces and "slicing" the 3D models into flat layers that could be translated into planar 2D (X and Y) movements to direct the machine to build or deposit one layer. The machine would then increment downward (increment relative Z axis motion) and spread a new layer of powder to build the next layer and

[1]The Role of the National Science Foundation in the Origin and Evolution of Additive Manufacturing in the United States, Institute for Defense Analysis, IDA, SCIENCE & TECHNOLOGY POLICY INSTITUTE, November 2013, Christopher L. Weber, Vanessa Peña, Maxwell K. Micali, Elmer Yglesias, Sally A. Rood, Justin A. Scott, Bhavya Lal, Approved for public release; IDA Paper P-5091, Log: H 13-001626, https://www.ida.org/ ~ /media/Corporate/Files/Publications/STPIPubs/ida-p-5091.ashx, (accessed December 19, 2016).

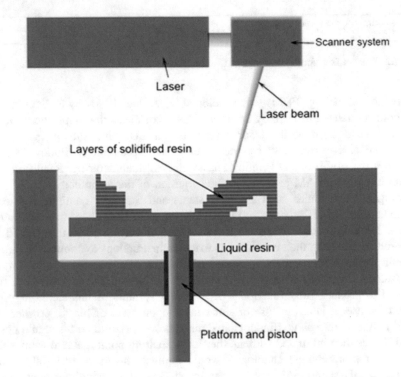

Fig. 7.3 Stereolithography apparatus schematic "Stereolithography apparatus," https://upload.wikimedia.org/wikipedia/commons/1/1e/Stereolithography_apparatus.jpg[2]

repeat the process. This is also known as 2 ½ axis (or 2 ½ D) fabrication as all of the Z motion is realized in incremental Z axis steps.

Today an ever-increasing variety of materials is being developed and printed from ceramics, to composites, to biomaterials such as living cells. In some cases products are being made to replace those conventionally made using metal such as jigs and fixtures for manufacturing. New materials developed specifically for 3D printing are being realized across the globe every week. Another trajectory for this technology is that of the personal 3D printer. These systems are at a price point attractive to individuals and have found early adoption in educational and recreational use. Quality and functionality is increasing rapidly with name brand companies entering the marketplace. Advances in 3D printing plastics and in adoption, application, and entry into the manufacturing value chain, serve as a model for those advances being developed for metals.

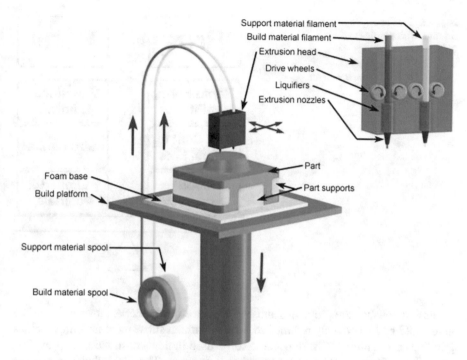

Fig. 7.4 Fused deposition modeling[3]

Prior to and during the development of rapid prototyping and 3D printing of plastics a number of other technologies were seeing steady advances as well. Let us review these and see how these technologies evolved, in parallel with 3D printing of plastics, into the AM systems we use today.

7.2   Weld Cladding and 3D Weld Metal Buildup

Weld cladding has been around almost as long as welding. Historically applied as a manual process using flame or arc torches, a buildup of weld filler materials upon a substrate part could be used for fabrication of features, repair, renewal, or upgrade of components. Figure 7.6 shows a pipe clad by welding with a delivery head and inert shield shown in position. Often cladding is applied to offer benefits for wear or corrosion resistance. An example is the weld clad repair of a backhoe bucket tooth. A worn down backhoe bucket tooth can be rebuilt to shape by successive weld beads of hard abrasive resistant metal, placed one next to the other, layer upon layer, to reform the original shape of the part. Repair or renewal offers an opportunity to upgrade the hard alloy of metal weld filler to improve the quality of the

[3]Courtesy of CustomPartNet Inc., reproduced with permission.

Fig. 7.5 Pros and cons of 3D printing with plastics and polymers

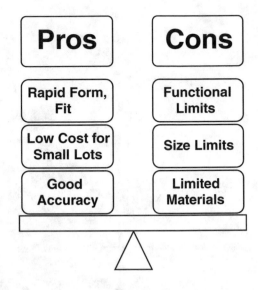

repair and enhance the properties and performance in service. A repair made in the field could be immediately returned to service without subsequent finishing such as grinding or machining. Cladding can also be used in new construction to provide a corrosion resistant coating or wear resistant features. The weld cladding process relies on a base part and imparts an enhanced function (Fig. 7.7) usually localized to a specific region of the part and often used to enhance or repair a part surface. In an historical example, in the 1970s, Thyssen a West Germany company manufactured a 19-ft-diameter × 34-ft-long cylindrical pressure vessel from ferrite materials by depositing multiple submerged arc welds against a consumable mandrel (Kapustka 2014; McAninch 1991).

Limitations to weld cladding using arc heat sources include the large molten pool, usually limited to flat position deposition and the large amount of total heat input resulting in heat buildup, and the potential for distortion and residual stress within the part. As a result, weld cladding was often limited to large parts that could be articulated to the flat position and were able to withstand the thermal/mechanical stress and induced distortion (Kovacevic 1999; Brandl 2010).

Other repair applications include resurfacing railcar wheels, jet turbine vanes, and rebuilding worn marine shafts and other wear surfaces. After buildup with weld deposit, a finishing operation is often used to achieve the surface and dimensional specifications. Automated versions of the process can replace the operator with mechanized motion or CNC motion control and can use wire feed for filler, metal powder filler, or even strip filler to enable higher deposition rates.

The integration of lasers and plasma arc welding systems, powder or wire delivery, and inert gas chambers has been demonstrated. Induction preheating, pulsed laser, pulsed micro-GTAW and variable polarity plasma arc control may be

Fig. 7.6 Welded clad pipe with delivery head and inert shield in position[4]

Fig. 7.7 Cladding used to enhance or repair a base part

used to limit heat input and tailor the resultant microstructure. Laser scanning and vision-based controls are also being used. Three to eight axes of fully coordinated motion is available.

7.3 Laser Cladding

Lasers have made significant inroads in weld cladding operations as they are easily integrated into the production environment, and are now cost effective and industrially hardened to operate in the harsh condition of the shop floor. Laser-deposited

---

[4]*Source* Geoff Lipnevicius, "Robotic Applications for Cladding and Hardfacing," Fabricating & Metalworking Magazine, November 29, 2011. http://www.fabricatingandmetalworking.com/2011/11/robotic-applications-for-cladding-hardfacing. Reproduced with permission.

clad material can in some cases be deposited faster and more accurately than arc-based systems and in many cases offers benefits in the microstructure, metallurgy, and quality of the final deposit. A more accurate deposit wastes less filler material and can be removed faster when employing machining or grinding to final dimension. The more highly focused heat source can reduce heat input resulting in a more highly refined microstructure, reduced thermal distortions, and residual stresses. Strict control of penetration into the base material can control the percent dilution of base material to cladding material resulting in a tighter control of the metallurgical properties of the deposit. Therefore, filler alloys and the clad process must be carefully controlled to achieve the desired properties of the resulting clad for a given base metal. The metallurgical engineer must be cognizant of any modification or changes in procedure as small changes can result in large effects in the resulting microstructure, defect morphology or residual stress in the final part. A good reference for laser cladding is (Palmer and Milewski 2011). Weld cladding made slow but steady evolutionary progress during the rapid growth years of CAD/CAM, rapid prototyping and lasers as a mainstream industrial proven and certifiable process. Laser heat sources have replaced arc and plasma in many applications as lasers deposit less heat input with faster travel speeds that can result in less distortion or shrinkage stresses. The small molten pool may be more easily articulated to allow out of flat position deposition. In addition, there are numerous benefits to the chemistry and microstructure of the laser clad deposit. Laser-based systems cost more than arc-based systems and are more complex. Laser safety hazards must be controlled and can limit usage in on-site locations.

The evolution of weld clad processes into directed energy deposition AM processing is primarily based upon the use of computer models to generate 3D deposition paths, the use of multi-axis control and the evolution of laser heads with various powder and wire feed configurations. Modern hybrid AM machines combined with CNC lathes and mills have demonstrated both cladding and machine finishing of the clad in a single workstation. We will elaborate later when discussing AM directed energy deposition (DED) and hybrid systems.

7.4 Powder Metallurgy

Powder metallurgy (PM) objects are formed by compaction using punch and die tooling and sintering of metal powder and can have advantages over those produced by melting, alloying, and fusing. Powdered materials, chemically incompatible when melted, can be pressed together to form a self-supporting *green part*. This part is then heated in an inert atmosphere furnace and sintered to form a useful object. The process can be cost-effective in producing large quantities of near net shaped metal objects that may require fewer secondary or finishing operations when compared to alternative means of manufacture. Energy and time savings may be realized using the process. Material utilization can be as high as 95% making it attractive for certain costly material applications. Materials, otherwise unable to be combined using melt processing such as tungsten carbide with steel, can be used to produce metal matrix cutting tools or hard materials may be combined into components such as cutting tool inserts. Other composite materials may be formed by mixing metal powder with nonmetallic powders such as graphite to form electrical motor contacts and brushes. Porous materials such as bronze bearings that hold lubricating oils within interconnected porosity have seen wide application. Parts with complex cross-sectional shapes, such as automotive transmission synchronizer sleeves, requiring close geometrical tolerances can be fabricated and may reduce or eliminate the need for machining. High strength may be achieved, but sintered metal parts often suffer from low elongation properties due to voids and imperfections within the bulk material. A typical powder metallurgy part process flow is shown in Fig. 7.8 with the pros and cons listed in Fig. 7.9.

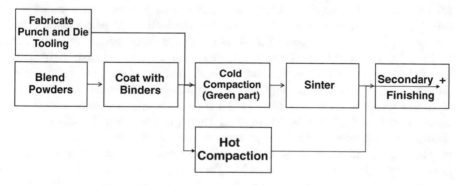

Fig. 7.8 Powder metallurgy process flow

Fig. 7.9 Pro and Cons of powder metallurgy products

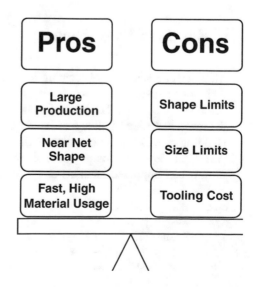

Significant technology development has occurred in the production of powder materials, powder characterization, and the characterization of sintered metal materials and parts. Much of what has been learned is directly or indirectly applicable to AM sintering or AM fusion processing and is helping to create new industry standards.[5]

Limitations of powder metallurgy processing include the high cost of punch and die tooling and the high capital equipment cost of secondary processing equipment such as hydraulic presses, furnaces, and hot isostatic presses. These costs must be amortized over the production life of a part and typically require tens of thousands of parts to recover these upfront costs. Parts are often limited to smaller sizes due to limitations in pressing equipment capacities and tooling costs. Part design limitations include size, aspect ratio, and limits to features such as sharp edges, bevels, chamfers, and sharp corners. Reentrant features that would prevent the removal of the pressed part from the mold, such as grooves, reverse tapers, and lateral holes cannot be formed. Automotive components such as pulleys and gears are ideally suited due to their size, geometry, tolerances, mechanical property, and service requirements. The shape of powder particles used in PM processing is often angular or irregular assisting in packing density and pressed green shape strength as opposed to spherical such as is used for AM processing. A good reference is provided by the Powder Metallurgy Review, "Introduction to Powder Metallurgy".[6]

AM of metals and PM may evolve to be complementary processes because the biggest drawback for PM is the cost of mold making and the need to justify the high upfront costs of tooling with large production runs. New applications using AM may help speed the process of prototype mold development by producing short run tooling for small punches, dies, and molds. Tooling cost reduction realized by using AM may lower the investment recovery level of PM making it economically viable for smaller production runs.

7.5 Key Take Away Points

- Much can be learned by becoming familiar with the existing technologies from which AM metal has evolved. Rapid prototyping with plastics and polymers launched the use STL models and layer-wise deposition. Weld cladding demonstrated metal buildup of coatings and shapes while laser welding and cladding provided a better understanding of the metallurgy of rapidly solidified AM metal deposits. The powder metallurgy industry continues to lead the way in new and improved methods of powder production.

---

[5]ASTM International Committee F42 on Additive Manufacturing Technologies, https://www.astm.org/COMMITTEE/F42.htm, (accessed May 14, 2016).

[6]Powder Metallurgy Review, provides access to "Introduction to Powder Metallurgy": A free 14 page guide, www.ipmd.net, (accessed March 20, 2015).

- The origin technologies of AM metal provide a rich source of information with engineering studies and applications relevant to AM metal processing such as metal powder and production facility safety. Although more mature, these technologies continue to evolve in parallel with AM metal. Many have been at a full production technology and manufacturing readiness levels (TRLs and MRLs) for decades.
- Technical experts and technical professional societies associated with these technology represent and large body of relevant knowledge. Knowing who these people are and where to find these resources is valuable to those newcomers to AM metal processing.

Chapter 8
Current System Configurations

Abstract System configurations for additive manufacturing metal are most often described and differentiated by the heat source used, such as laser, arc or electron beam, how the feedstock is delivered, the type of feedstock used, such as wire or powder, or the size of the part produced ranging from in meters to millimeters. It is useful to understand the basic system configurations as they all feature different attributes and capabilities. There are advantages and limitations to each and it is important to the user to understand these variations to make an informed decision as per which is the best for their needs. This chapter provides a technical description of the basic functions and features of each type of system. In addition other hybrid process that begins with a 3D computer model and results in a metal part are also described as these can in some cases be a competitive option to those systems that go from model directly to metal. Processes that exist on the border of the more common definition of AM metal, such as those that produce parts at the micrometer and nanometer scale, are also introduced.

What do you get when you combine lasers with computers, solid models and CNC robots? One answer is an AM metal printer. In this chapter we discuss current AM systems that begin with 3D solid models, utilize computer motion control and focus on high energy heat sources to fuse metal into solid metal objects. We talk about which systems use lasers, which use electron beams or electric arcs, and why some systems use powders, while others use metal wire as feed material. We talk about what they have in common, the advantages and disadvantages of each. We also introduce other additive manufacturing processes, not based on high energy heat sources that fall under the category of 3D metal printing. Examples are provided to compare 3D printing with conventional processing. Additional examples taken from industry, published reports and Web content are used to highlight where each system technology is today.

What are the different types of AM metal systems? (Fig. 8.1) How does each method start with a model and end up with a part? What are the pros and cons of

© Springer International Publishing AG 2017

J.O. Milewski, *Additive Manufacturing of Metals*, Springer Series in Materials Science 258, DOI 10.1007/978-3-319-58205-4_8

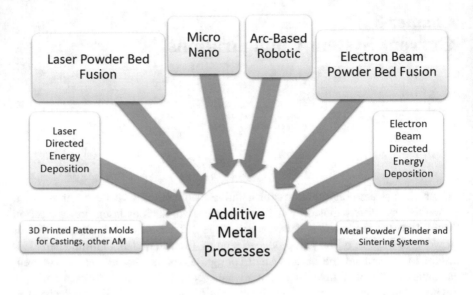

Fig. 8.1 Additive manufacturing metal processes

each process? Which is best for you? Depending on the material and process, the end product may be substantially different. After reading this chapter the informed user will be better equipped to choose the right process based on the requirements of the design and end use of the part.

The readers should keep in mind there are many variations of the fused metal deposition technologies discussed here as provided by a wide variety of vendors. Their specific methods may handle these technical challenges differently although much of the technical detail of how these challenges are handled by the machine, process or software may not be evident until you buy the machine, take the training and start building parts. As such, these discussions will be kept generic and not focus on a specific vendor or vendor technology when discussing common challenges. Later in the book I will provide a few examples and links to specific vendor and organization Web pages that describe unique or novel methods, capabilities, or demonstrations. I seek to engage and open the discussion and exposure of AM technology to a wider audience and I wish all vendors and organizations success in carving out a unique value position within this rapidly expanding field. Leading vendors and their vendor specific process names are listed in Table 8.1.

Two general AM methods for rapid prototyping metal emerged about 20 years ago. ISO/ASTM 52900,[1] now defines them as Powder Bed Fusion (PBF) and Directed Energy Deposition (DED). Within this text we clarify the uses within the

---

[1]ISO/ASTM 52900, Additive manufacturing—General principles—Terminology, http://www.iso.org/iso/catalogue_detail.htm?csnumber=69669, (accessed April 18, 2016).

Table 8.1 AM metal equipment manufacturers and their specific process names

Process	Process name	Manufacturer	ASTM category
DMLS	Direct Metal Laser Sintering	EOS	PBF-L
SLM	Selective Laser Melting	SLM Solutions	PBF-L
DMP	Direct Metal Printing	3D Systems	PBF-L
LaserCUSING®	LaserCusing	Concept Laser	PBF-L
EBM®	Electron Beam Melting	Arcam AB	PBF-EB
EBAM™	Electron Beam Additive Manufacturing	Sciaky Inc.	DED-EB
LENS®	LENS	Optomec	DED-L
DMD®	Direct Metal Deposition	DM3D Technology LLC	DED-L

context of AM metal processing by adding a designation of the heat source used, such as L for laser beam (DED-L) or EB for electron beam (DED-EB).

PBF scans a high power laser or electron beam along a prescribed path to fuse a pattern, derived from by a sliced layer of an STL model, into a bed of metal powder. The powder bed is incrementally moved downward and another layer of powder is added by a recoating blade or roller. The process is repeated with the high energy beam fusing the next slice from the model, followed by another incremental downward motion and recoating a layer of powder. The process of recoating, fusing and downward movement continues until the part is complete. PBF processes using laser beams (PBF-L) are widely referred to in the literature by names such as direct metal laser sintering (DMLS), selective laser melting (SLM) or selective laser sintering (SLS). The PBF process using an electron beam (PBF-EB) is also known as EBM or electron beam melting. In our generic discussion we will use the terms PBF-L and PBF-EB. Leading vendors and their vendor specific process names are listed in Table 8.1 to assist the reader when searching the Web.

DED involves delivering powder or wire into the focal spot or molten pool created by a laser, electron beam or plasma arc directed at a part surface, completely melting and fusing the filler and translating this deposit to build up a part as directed by a 3D deposition path. DED processes using laser beams (DED-L) are widely referred to in the literature by names such as laser engineered net shape (LENS), direct metal deposition (DMD) and, laser metal deposition (LMD). The DED process using an electron beam (DED-EB) is also known as electron beam freeform fabrication (EBF3) and electron beam additive manufacturing (EBAM). Plasma arc based systems will be referred to as PA-DED. In our generic discussion we will use the terms DED-L and DED-EB. First we discuss the advantages and disadvantages of PBF-L, the most widely applied of these processes.

8.1   Laser Beam Powder Bed Fusion Systems

The general principle of selective laser sintering, as applied to metal as in PBF-L, is shown in the schematic of Fig. 8.2. The laser beam is directed at a bed of powder to fuse a layer defined by the cross sectional area of the sliced part model and a scan path (Fig. 8.3) of overlapping weld beads. The powder bed and part are then incrementally dropped and recoated by a roller or blade spreading a new layer of powder to allow the fusion of the next and successive layers of powder to form the part. It is important to note the powder layer thickness is greater than the fused deposit layer thickness. The depth of penetration is greater than the deposit layer thickness and can often penetrate three or more layers in depth to more fully fuse the deposit. PBF-L has evolved considerably over the years to the point where a near 100% fully dense metal part can be fabricated directly from 3D computer models. Common engineering alloys based upon steel, nickel, titanium, cobalt chrome molybdenum (CoCrMo), metal matrix composite materials and other specialty metals are used in PBF-L. Build speed, dimensional accuracy, deposition density and surface finish improvements have improved steadily. The manufacturers of this equipment continue to design, build and sell larger and more capable equipment. Precompetitive research continues in the universities and corporate research labs, but as we will discuss later, consortiums with in-kind funding from government and industrial partners are becoming widespread.

Partnerships between machine sellers, software vendors, powder manufacturers and end users are paving the way for adoption in a wide range of industrial applications and business sectors. In some cases specialty components are making their way into production environments for small lot size or custom components. Significant inroads have been made into the medical, dental and aerospace sectors, as was shown later in Chap. 2, featuring novel designs and new and interesting applications.

It can be difficult as a Web based observer to separate the proof of concept demonstrations, from actual functional prototype testing to the real production examples and money makers. There are some emerging applications that could be considered or potentially disruptive as in the case of dental crowns and implants. Given the cost of equipment ranging from hundreds of thousands of dollars to millions of dollars, most of the work is still being done by highly skilled and equipped engineers in corporate R&D and university lab settings or by service providers able to make these up front investments. There are a growing number of private AM fabricators who have made the leap into the service sector by purchasing the latest AM metal systems and offering AM fabrication through Web based services. It is only a matter of time before there is an AM metal capability in a city near you at an affordable cost.

But first, let's step back and consider some of the current features common to most systems, as well as advantages and drawbacks of the PBF-L process and see

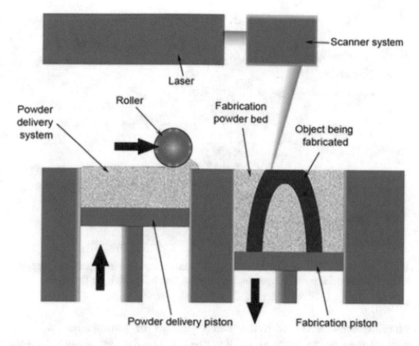

Fig. 8.2 Selective laser sintering process. "Selective laser melting system schematic," https://
upload.wikimedia.org/wikipedia/commons/3/33/Selective_laser_melting_system_schematic.jpg[2]

where the entry level user can access the technology. Later in the book we will
introduce the other PBF-EB and DED systems, and compare and discuss where
they all stand within the larger picture of AM.

8.1.1   Advantages of PBF-L

A big advantage of the PBF processes is the wide range of CAD software that can
be used to generate STL files for these machines. The wide availability of STL file
editing software allows fixing, editing, slicing and preparation for 3D printing.
The STL files may be oriented and duplicated as required to utilize the build
volume efficiently. Support structure design may be required depending on the
geometry of the object to be built as unsupported material can warp or distort if not
anchored by the support. As will be discussed in more detail later support structures
may also serve as heat sinks and prevent movement or disorientation of small
feature during the spreading of powder layers. Model slicing, as with plastic AM

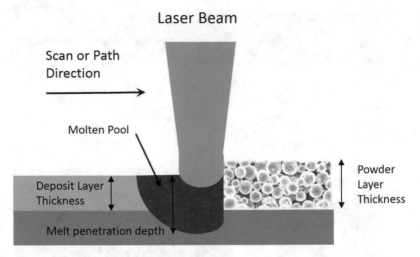

Fig. 8.3 Laser scanning showing a melt depth penetrating into the previous deposit, and comparing the as-spread powder layer thickness to the fused deposit layer thickness

machines, creates layers with the hatch patterns or scan paths and machine instructions required to deposit each layer to produce the part. Figure 8.4 shows a computer model of a typical support structure shown in red and the part shown in gray.

Recommended machine parameters are often available from vendors for a subset of well-known materials, but often at additional cost. User-defined parameters may be developed, but detailed knowledge and experience with the process is required to select scan speeds, Z height steps and path offsets to assure a uniform deposition, full density and to attain the desired material properties. In time, designers and makers will get comfortable with the process as has already happened with 3D plastic printers. In time the learning curve for metals will become less steep, the price of materials will decrease and the penalty of learning the hard way through mistakes will decrease. With experience, realizing complex designs in metal will effectively be but a click away for a wide range of materials.

Laser scanning optics relies on magnetically driven mirrors using galvanometers. This method is most commonly used to allow rapid movement of the beam impingement location within the build volume. This method avoids the need to articulate the mass of a laser head's final focusing optics, such as with DED-L, to achieve accurate X- and Y-axis beam positioning at rapid speeds. In comparison with DED-L, rapid movement of the entire mass of a laser head is subject to delays during hard acceleration or deceleration and requires a rigid and massive mechanical system to maintain the accuracies and speeds required. Therefore, the simplicity offered by scanning optics, where only mirrors are moved, is an advantage.

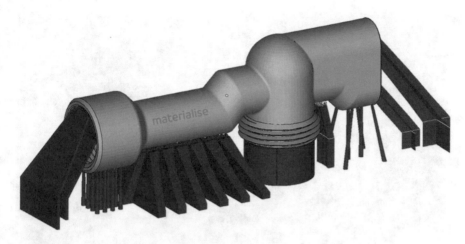

Fig. 8.4 Solid model with support structure shown in *red*[3]

Powder bed methods offer the opportunity to build multiple instances of the same part all at once.[4] In addition, multiple instances of different parts may be built at the same time. Software for optimizing the positioning of parts within the build volume, with various virtual objects, all to be built at once, is already being offered. In another example an external reamer tool fabricated by selective laser sintering (Fig. 8.5) features a rib structure inside the tool reducing weight by one half. The reduced inertia of the tool enables faster machining and higher precision.[5]

Recent process enhancements include increased processing speed by heating the powder and higher purity inert gas supplies for reactive metals used in critical applications. Inert gas is also used to accelerate the cooling after completion of the build cycle. Many of these processes operate in a fully unattended mode, allowing round the clock processing. Many vendors offer remote viewing and real-time process monitoring.

Unique metal part shapes can be fabricated that cannot be fabricated by conventional means. Structures with complex shells, internal lattice structures, internal

[3]Courtesy of Materialise, reproduced with permission.

[4]Concept Laser press release, Report: Mapal relies on additive manufacturing for QTD-series insert drills, July 6, 2015, http://www.concept-laser.de/en/news.html?tx_btnews_anzeige[anzeige] =98&tx_btnews_anzeige[action]=show&tx_btnews_anzeige[controller]=Anzeige&cHash=9fb996 72e9eac2b5e43e11fbb4e65198, (accessed August 14, 2015).

[5]Weight optimized external reamer, Mapal, http://www.mapal.com/en/news/innovations/laser-sintered-external-reamers/?l=2&cHash=a80b7bbe9ac848c98ad82794e4088bbd, (accessed January 29, 2017).

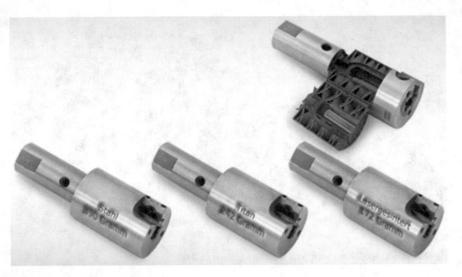

Fig. 8.5 Low inertia external reamer tool bit fabricated by selective laser sintering[6]

cooling channels, or complex superstructures have been demonstrated. Complex features such as these can minimize the use of metal, optimize strength, or extend functionality. Building in functional features can optimize gas or fluid flow, cooling, or other thermal or mechanical properties. Complex internal passageways can be formed provided that powder trapped during the build cycle and any required supports can be removed during post-process finishing operations.

High-performance materials, composites, and even ceramics have been demonstrated and offer the promise of hybrid, custom components made economically from materials previously unavailable. AM designs may combine what historically were a number of parts requiring joints, assembly, and fasteners into a single functional component.

A big advantage of solid freeform design and AM is the freedom from the constraints of commercial shapes and the reliance on easily fabricated materials. A reduced reliance and investment up front on commercial process equipment and tooling may in certain instances be realized. As we will discuss in more detail later, a total life cycle approach from raw metal ore extraction to part replacement, removal from service and recycling will help to identify the real economic benefits of these AM processes. Five advantages of PBF-L are shown in Fig. 8.6.

---

[6]Courtesy of Mapal, reproduced with permission.

Fig. 8.6 Advantages of laser
powder bed fusion

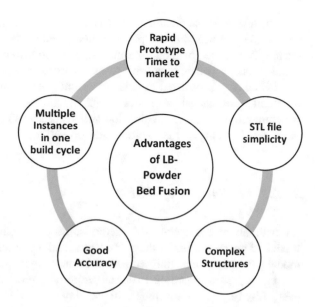

An entirely new paradigm for freeform design will eventually take hold allowing computer algorithms to optimize both designs and processing schedules to build parts with the best materials, least energy, lowest cost and most rapid response times. However, maximizing the benefits offered by AM design is currently limited to the hundreds of AM processing variables and limitations of human designers to optimize designs and parameters. Trial and error development or rule of thumb decisions are made based on limited experience or sparsely populated datasets.

Repair operations have been demonstrated for high-value components by removing the area to be repaired leaving a planar surface that can be held in a fixture and oriented as co-planar to the build surface within the build volume. This orientation allows typical 2 ½ D layered deposition to proceed from that point on remanufacturing the features above that region. This may provide the opportunity for remanufacturing improved or enhanced features using higher performance materials resulting in either better performance or longer life of the component, although DED-L is better suited to these applications. Precise repositioning of the part within the powder bed and realignment with the recoating blade may in practice limit these applications.

Post AM process operations such as heat treatments may be used to transform a near shape part to a finished part enhancing the as-deposited properties or performance. Powder removal and cleaning may be followed by surface finishing operations such as peening, polishing, or coating. Furnace heat treatments or HIP processing may be utilized to reduce thermal stresses, homogenize microstructures or modify mechanical properties. CNC machining may be required for support structure removal and full realization and accuracy of certain features. We provide more details and revisit post-processing operations later in the book.

8.1.2   Limitations of PBF-L

As with all of the metal AM methods, process complexity remains an issue. Increased understanding of the best designs and the necessary process control, from model generation to the finished part, is required to realize the full potential of these processes. Issues regarding material properties, product consistency, process repeatability (e.g., same machine different day or moving from one batch of powder to another) and process transportability (different machine, at a different location, using the same parameters) need to be fully addressed to gain the confidence required for material and process standardization and certification when used in critical applications. The major corporate players, government consortiums, and standards organizations realize this and are making progress to identify and resolve these issues. We discuss this in more detail later in the book.

Powder bed fusion processing, utilizing the sintering or melting of metal powder (Fig. 8.7), can achieve as-deposited densities of up to 100%. Controlling the melt pool size, powder layer thickness, laser power and travel velocity of the melt pool $V_{melt pool}$, and hatch spacing or scan line offset (Fig. 8.8) is critical to fully melt and fuse the deposit into adjacent layers and fully penetrate into previous layers of deposit for a given hatch spacing and layer height.[7] Figure 8.9 shows unfused regions of powder of a type that can result from a process disturbance or an inadequate parameter selection. Other process limitations are shown in Fig. 8.10

---

[7]The Effects of Processing Parameters on Defect Regularity in Ti-6Al-4 V Parts Fabricated By Selective Laser Melting and Electron Beam Melting, Haijun Gong, Khalid Rafi, Thomas Starr, Brent Stucker.

SFF, http://sffsymposium.engr.utexas.edu/Manuscripts/2013/2013-33-Gong.pdf, (accessed May 14, 2016).

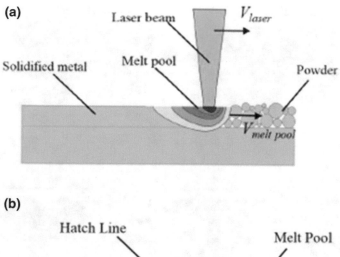

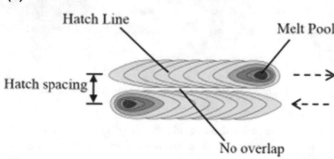

Fig. 8.7 Powder bed fusion relies on hatch spacing to assure overlap of weld beads[8]

while a more complete discussion of PBF process related defects and detection is provided later in the book.

It took a couple of decades of powder and PBF-L process development to attain the goal of 100% density for certain materials. More experience and the building of parameter data bases is needed to gain an acceptable level of confidence for a full range of AM deposited materials. Flaws in bulk metal and finished components are a way of life for any material processing operation, but knowing what to expect and what is allowable will require a concerted effort over the next decade. Discontinuities, *flaw* content and *anisotropy* within the microstructure of AM parts will be revisited later in the book.

[8]*Source* Haijun Gong, Khalid Rafi, Thomas Starr, Brent Stucker, "The Effects of Processing Parameters on Defect Regularity in Ti-6Al-4 V Parts Fabricated By Selective Laser Melting and Electron Beam Melting", D.L. Bourell, et al., eds., Austin TX (2013–33) pp. 424–439. Reproduced with permission.

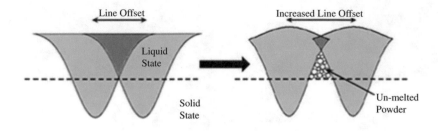

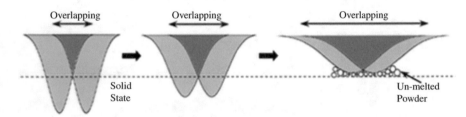

Fig. 8.8 Adjacent Melt Tracks Must Penetrate Into the Layers below to Achieve Full Fusion[9]

As introduced earlier in the discussion regarding the properties of AM metal and may display micro-porosity, leading to decreased properties and performance associated with fatigue life, elongation, impact toughness, creep, rupture, or loss of strength and ductility. The goals of 100% density for all materials, under all deposition conditions, will conflict with the goals to speed deposition rates. Rapid solidification rates may result in metastable microstructures and material textures that are detrimental in the as-deposited conditions that are detrimental and will require HT or HIP post-processing.

Size or capacity limitations exist for all processing equipment with good reason. A watchmaker needs a different lathe than one specifically designed to machine truck axles. Equipment cost and accuracy come into play as well. Therefore, the dream of one 3D printer for all objects and all materials is yet to be realized and may never be. Large machines built specifically for one task, such as 3D printing an automotive body or the backbone of a jet fighter is being proposed. Commercially available professional PBF-L systems are currently limited to building components of a maximum size on the order of a 400–500 mm.

[9]*Source* Haijun Gong, Khalid Rafi, Thomas Starr, Brent Stucker, "The Effects of Processing Parameters on Defect Regularity in Ti-6Al-4 V Parts Fabricated By Selective Laser Melting and Electron Beam Melting", D.L. Bourell, et al., eds., Austin TX (2013) pp. 424–439. Reproduced with permission.

Fig. 8.9 Regions of Unfused Powder[10]

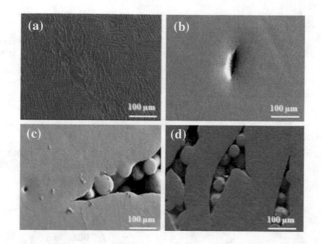

A larger build volume will not only require more material for the actual part but also generate more material requiring reuse or recycling. Powder bed system raw material requirements will scale directly with the build volume, while DED powder feed systems will maintain approximately the same fusion efficiency. Wire feed systems feature the greatest efficiencies, approaching 100%.

The time required to spread or recoat a layer of powder and delays, such as those encountered during preheating and cooling the powder bed, will scale with build volume size. Building smaller parts within a larger build volume will see a significant drop in efficiency as the beam on time will scale with part volume while the recoat time will scale with build chamber dimension and build height. A customer choosing a service provider with a much larger build chamber than required for their part may end up paying for unneeded capacity, materials, time and resources. Machine vendors are continuing to reduce recoat times using bi-directional, circular or other innovations. The precision of the recoating process requires precise setup and is subject to process disturbances which may result in uneven buildup or a recoat blade crash and process interruption. Process restart may be possible but difficult.

Dependency on conversion to STL format limits the ability to carry design information to the machine. As stated earlier, these limitations are to some degree being addressed in the development of the 3MF file format. Feature-based information and sequences, such as those used in CNC machines, is lost when the part is represented as an STL surface model sliced in a single orientation.

[10]*Source* MSA, Materials Sciences and Applications, Vol.3 No.5(2012), Article ID: 19181, 6 pages DOI:10.4236/msa.2012.35038, Effect of Melt Scan Rate on Microstructure and Macrostructure for Electron Beam Melting of Ti-6Al-4 V. Reproduced with permission.

Fig. 8.10 Potential
disadvantages of laser powder
bed fusion

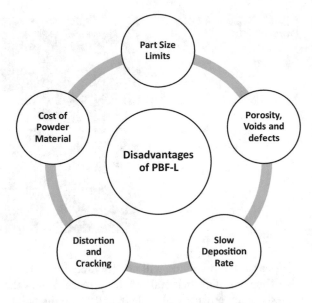

The weight and cost of powder to fill a large build volume may in itself be the limiting factor to scale this method to large monolithic objects. Real-time powder collection and reuse is currently being featured on some systems to improve the powder handling efficiency and safety of the entire build cycle as well as modular powder handling systems and an increasing level of automation to assist in the building of increasingly larger objects. Software to design segmented components to be assembled and joined after building is currently available to help overcome small build chamber size limitations.

The size of any part is limited by the dimensions and capacity of the build box and support environment. Sufficient metal powder must be used to attain the desired build height; therefore, a large volume of powder is needed for the process that does not become part of the object. Although material utilization can be high by sieving and recycling un-melted powder, this relies on purchasing and handling large quantities of powder and accepting the associated costs, difficulties and hazards of those operations. As an example, building a spherical titanium shape of 20 cm in cubic build volume of 20 cm per side will require 8000 cm^3 of powder weighing 36 kg or nearly 80 lbs. Scaling to build a 40 cm sphere would require 288 kg of titanium powder or nearly 640 lbs. Never mind the build time or the actual weight of the part; the cubic scaling of powder required to achieve large part dimensions rapidly makes current machine configurations unrealistic for very large parts.

Specialty AM powders offering high purity, chemically clean, consistent particle sizes and shapes, are costly and in some cases in limited supply. Existing supplies of commercially available metal powders, optimized for conventional powder metallurgy processes, such as pressing, sintering and spraying, are plentiful but have not been optimized for PBF-L. In contrast, powders used for DED processing often have less stringent requirements and are available in a wider variety of alloys without the premium price.

Recent progress has been made in the production of metal powders addressing cost, supply, morphology and safety, but a greater number of economically viable sources for AM metal powder must be established if these AM methods are to enter into widespread use for final part production.

Powder not fused into the parts and supports during a build may be reclaimed and sieved to remove partially fused lumps and mixed with new virgin powder. Material traceability for critical applications may be lost when *virgin powders* from one chemistry lot are mixed with *secondary powders* reclaimed from a previous build cycle. Research is ongoing to determine how often an AM specialty powder may be reused before changes to the morphology, chemistry, or particle size distribution render it unacceptable.

Dimensional accuracy is tied to the laser beam spot size, powder size, and part orientation. Larger spot sizes allow faster build rates but produce less accurate features. Large PBF build chambers require scanning the beam across a wider area plane with a corresponding need to change the focal conditions resulting in larger impingement angle of the beam to deposition surface. Changes in laser power alone can alter focal spot conditions requiring changes in multiple processing parameters to maintain the quality of the deposited materials. When laser optical scanning systems change the location of the laser focal spot in X, Y, and Z, the intensity profile of the laser spot energy distribution will change and can limit the extent of the laser scanning volume. All this adds to the complexity associated with process schedule planning in open or closed loop real-time control modes. Commercial systems with variable focal spot sizes are now being offered to address many of these conditions.

Surface condition and roughness can vary depending on powder morphology, build conditions, and part orientation within the build volume. These issues may be controlled to some extent with well-developed procedures, control of powder reuse, control of part placement, and orientation within the build volume, but these unknowns will always be part of the character of the current generation of PBF machines.

For all powder-based AM processes, free or loose powder particles must be cleaned off exterior or interior surfaces of the as-built part and may require finishing depending on the part application, adding to the number and type of post-process finishing steps.

Concerns associated with part orientation or location within a fully loaded build environment shared between customers may exist. Concerns for part quality or repeatability need to be addressed, as currently the same model sent to different vendors may result in parts with different dimensional and surface features. Will you pay extra to accommodate the risk of a failed build due to someone else's part design being built at the same time as yours? Will a full description of the build environment conditions be provided by the vendor to the customer for every part fabricated?

Geometric limitations such as maximum overhang angles, as an example 35 degrees, need to be maintained to minimize the use and subsequent removal of support structures.

DMLS or SLM is limited to a single material type within the powder bed. Functionally grading material properties as a function of part location or part feature is not an option. Repair applications made using this process may use a different powder composition, but this is not currently in widespread use. The process could be stopped, the part cleaned and machined and another set of features built off a new planar surface, but this would require stopping the process, cleaning the machine of old powder and starting over with new powder and a precisely reoriented part.

Environment, safety, and health issues associated with handling, storage and processing with metal powders need to be controlled. Powdered metals can be difficult to handle, store, and process. Improper storage can lead to oxidation or contamination sources or formation of other chemical compounds that present unique hazards. Finely divided powers can be pyrophoric and when ignited burn at temperature beyond the capabilities of hand held fire suppression equipment. Finely divided powders can easily become airborne and contaminate surfaces creating inhalation and ingestion hazards. Enclosed processing chambers, inert processing atmospheres, sealed storage containers, specialized vacuum, filtration equipment, and proper training are but a few of the engineering and administrative controls required for the safe use of powders.

Laser hazards must be well controlled. Industrial systems often provide *"Class I"* laser enclosures that contain the thermal and laser light hazards. High levels of formal training are required for the safe operation and repair of these systems as high powered industrial lasers, with invisible laser beams, are capable of projecting damaging laser energy or reflections great distances.

Change over from one powder type to another requires extensive chamber cleaning to prevent contamination of one alloy metal with another. Adverse metallurgical effects such as cracking, corrosion susceptibility or other effects may result from even small residual amounts of powder particles left within the build environment or cleaning systems.

Anisotropy, or variation of the grain structure and bulk properties, can occur within the parts as a function of material, build conditions and part orientation. Anisotropy, also known as microstructural *texture*, present within most metal components, is a result of material processing and is not necessarily undesirable but may be important to know for critical applications.

Post-processing using heat treatments, HIP cycles or a range of finishing operations, such as peening, chemical etch or plasma polishing, is required in some cases to achieve the desired properties, uniform microstructure, stress relief or the desired surface condition.

PBF system and service provider Web pages offer a wide range of details regarding systems, material data sheets, and industrial applications. Links to selected companies are provided in the AM Machine and Service Resource Links section of this book for further information.

8.2 Laser Beam Directed Energy Deposition Systems

Laser directed energy deposition (DED-L), also known as LENS or DMD, fuses metal filler into a 3D shape under computerized motion control, starting with a 3D solid model. While many of the advantages and disadvantages are shared with PBF-L methods, there are some significant differences. Rather than sintering a bed of powder as in PBF, DED fully melts metal powder delivered to a molten pool or focal zone by a powder delivery nozzle, as shown in the schematic of Fig. 8.11. The laser/powder delivery head is traversed followed by the melt pool, fusing the deposit onto the substrate as fully dense metal. Liquid phase sintering is not used in DED processing as the microstructure is fully evolved from a molten state. Full densification and fusion is assisted by mixing within the molten pool and does not specifically require remelting by subsequent layers to achieve full density. A large chamber DED-L system is shown in Fig. 8.12.

While PBF-L shares much of its origin to plastic prototyping technology, DED-L shares many process characteristics with laser weld cladding and in many ways is a hybrid combination of laser cladding and 5-axis laser welding. DED software can be more complex than PBF or laser cladding when relying on feature-based models and CNC tool path control versus strictly relying on planar slicing of STL models. A substrate plate or part is required upon which to begin the deposition. The substrate may or may not become part of the final part. In hybrid applications, DED may simply be required to add features to an existing base component or commercial feedstock shape. DED can deposit material on complex 3D surfaces (rather than simply flat surfaces and X, Y movement), utilizing 5-axis or more of simultaneous movement of both the laser deposition head and articulation of the substrate part.

DED-L may also be used to deposit planar layers in a 2 ½ D deposition starting with an STL file format, although support structures similar to PBF-L are not often used, limiting the deposit to shapes without difficult to form overhangs. There are numerous process variations, but for the sake of comparison we will make our arguments and comparisons with a DED machine with 5-axis of CNC control. Metal powder is delivered by inert gas in an inert chamber, to a point co-focal with that of the laser beam or to the location of the moving molten pool. Variations of 5-axis CAD/CAM/CNC software and CNC controllers are often used.

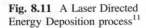

Fig. 8.11 A Laser Directed
Energy Deposition process[11]

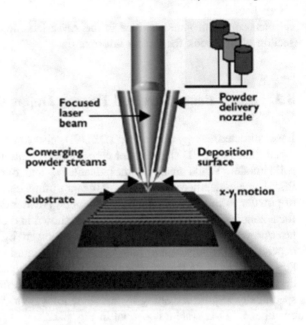

A primary difference between the various types of laser based powder fed DED systems is the laser head and powder delivery systems. A wide variety of laser/power heads are commercially available offering a wide range of capability. Knowledge of these differences and capabilities can help the end user to select the optimal configuration for their application.

While the basics of laser optics were discussed earlier in the book, the addition of powder feed takes the complexity to another level. As we recall, parameters of the lasers important to AM include as focal spot size, focal position and F# or convergence angle of the beam. Spatial intensity profile, beam power and axis of beam impingement also come into play. Powder delivery systems have an analogy to each of these laser parameters, including focal spot or *waist region*, convergence angle of the powder stream and focal or beam convergence location, all of which can affect the character of the deposit. The parameters of the powder delivery system such as powder feed rate, delivery gas flow rate, nozzle size, shape, location and powder impingement angle are key to consistent powder focus with respect to the laser focus at any tilt angle, speed or direction of movement. The laser parameters, combined with the powder parameters, make up the laser/powder interaction zone.

Gibson (2009, p. 243), provides a good introduction and illustration to powder nozzle configurations. The simplest configuration shown in Fig. 8.13a is a single wire feeder or powder nozzle (Fig. 8.13b) with a fixed relationship to the impinging

---

[11]*Source* Laser Engineered net shaping advances additive manufacturing and repair, Robert Mudge, Nick Wald, Weld J., 2007, 86, 44–48. Reproduced with permission.

Fig. 8.12 Model 557 Laser System (5′ *x*-axis by 5′ *y*-axis by 7′ *z*-axis)[12]

laser beam and the molten pool region. This is a common configuration used for laser cladding where metal powder is directed to feed the molten pool, to melt and form the deposit. In laser cladding of simple shapes with linear or rotational movement, large laser spot sizes, molten pools, high travel speeds and powder feeds can result in very high deposition rates. Focal position can be adjusted to affect penetration and the resulting percent dilution of the base material by the cladding material. Powder feed typically leads the path of the laser to melt the filler more efficiently.

The integration of lasers with powder feed into AM specific configurations has been developed in engineering R&D labs over the past couple of decades and is offered by various commercial vendors and system integrators. These systems utilize variations of co-focus/co-axial, laser/powder feeders into multiple powder feed streams with nozzles internal or external to the laser head. Optimization criteria for these designs include low mass to assist in articulation speed and small size to enable access to tight locations and to provide clearance for the laser head during tilt positioning to avoid existing part features. Multiple powder pathways (Fig. 8.13c, d) and powder feeders can enable the feed of multiple materials by switching from one material to another. Certain designs rely on the convergence of

---

[12]Photo courtesy of RPM Innovations, Inc., reproduced with permission.

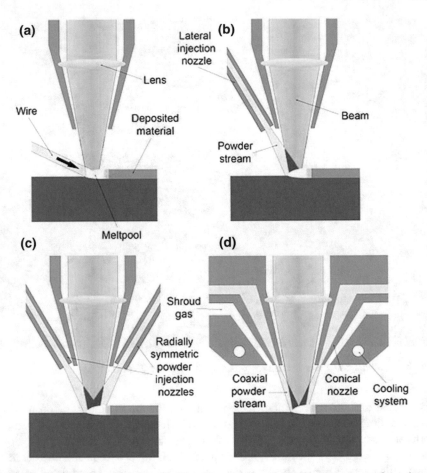

Fig. 8.13 Configurations of Laser Cladding Nozzles "Laser Cladding nozzle configurations," https://commons.wikimedia.org/wiki/File:Laser_Cladding_nozzle_configurations.jpg[13]

opposing powder streams to tightly focus the powder into the critical beam energy density region of the laser (Gibson 2009, p. 241). Ease of disassembly for cleaning, service and repair of the laser/powder head is also a consideration as are any embedded sensors or control devices.

Laser head and powder feed hardware can be large and bulky limiting rapid CNC motion and the range of axis movement. Deposition rates for laser DED are generally faster than for PBF but can be less accurate. Laser spot size in the

Fig. 8.14 Three powder
stream nozzle depositing on a
3D curved surface[14]

presence of the powder stream and melt pool size all affect deposition resolution. Some vendors employ dynamic sensing and control of the melt pool size to achieve a more uniform deposit.

In comparison, the movement of scanningmirrors to position the beam in PBF-L systems is significantly faster than CNC movement, but a direct comparison of deposition rate between PBF and DED would also need to take into account the speed of the recoat cycle. In DED-L, limitations associated with the articulation of the laser head mass may be offset by simultaneous movement of the part in relation to the head, but this solution would be limited when articulating large or massive parts. As you can see, there are many considerations and tradeoffs to take into account when comparing one system to another.

A single material feed location is often positioned at the leading edge of the molten pool or ahead of the beam impingement location to assist in the beam preheating and melting the powder or wire. As a result, this single material feed orientation must be preserved while following complex deposition paths requiring additional articulation of the material feed mechanism in relation to the deposition path and molten pool. A three powder feed stream nozzle configuration is shown in Fig. 8.14 depositing on a complex curved surface. Other specialty laser/powder heads are offered commercially, such as those used to clad internal bores of cylinders. Hybrid systems can feature fast deposition or fine detail deposition laser heads

8.2.1 Advantages of DED-L

Advantages of DED-L are shown in Fig. 8.15. Multiple powder or wire feeders may deliver different powder to the molten pool to allow a change in material composition during the deposition process, thus allowing a functionally graded deposit of metal. Switching between powder feeders allows the deposition of

---

[14]Courtesy of TRUMPF, reproduced with permission.

Fig. 8.15 Advantages of
Laser Directed Energy
Deposition

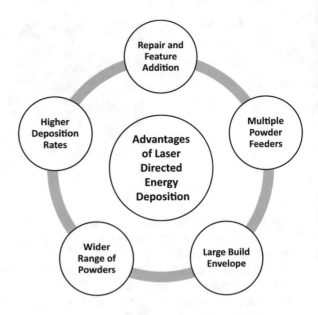

Fig. 8.16 Impeller Repaired
by LENS 850R System[15]

different features using different materials. Part repairs are facilitated by a wide
range of access to the part. An impeller part repaired by LENS DED-L process is
shown in Fig. 8.16.

A large and flexible build envelope is possible without the dimensional limitations of a powder bed. While laser cladding systems can deposit materials outside
the confines of a controlled atmosphere chamber, DED-L is most often performed

---

[15]Photo courtesy of Optomec (reproduced with permission); LENS is a trademark of Sandia
National Labs.

within a high purity inert glove box that contains powder hazards and restricts powder contamination into the factory environment. A high purity inert atmosphere can be maintained at oxygen and moisture levels of below 10 ppm (parts per million) using dry trains and other gas purification systems. Custom-sized chambers may be configured for building larger parts to provide a fully enclosed Class I laser containment enclosure as well as offer the possibility to recover and recycle unfused powder.

The ability to turn the powder feed off during a DED-L build offers the opportunity to use less virgin powder, glaze surfaces with a defocused or oblique laser impingement, and drill or clear holes or passageways using changes in peak power, focal or laser orientation. A system to rapidly switch the powder on and off has been developed at the Fraunhofer Institute in Germany. The ability to change the powder delivery gas offers the opportunity to control surface chemistry conditions, such as with nitriding.

A DED-L part is not buried within a powder bed during building and may be measured or interrogated during a build using non-contact or contact metrology methods to determine dimension or thermal conditions of the build, and control or modify the build schedule as appropriate. Hybrid systems incorporating AM, SM and metrology have been developed and demonstrated.

Additional control of positioning offered by CNC and multi-axis articulation during the build sequence allows feature based deposition and greater degrees of freedom in path planning and process control. As an example, distortion resulting from the deposition of one feature may be offset by building a mirror image feature across a build plane to offset and accommodate distortion and stresses of one feature versus the other during a build sequence. In comparison with PBF-L bed systems, distortion offset may be accommodated by the controller in the Z direction only. Otherwise software compensation of X or Y dimensions must be made to the original CAD model. DED offers the potential for shrinkage compensation by offsetting distortion and cancelation of opposing shrinkage forces and bending stresses rather than relying on software only.

Feature based parametric design software, such as that currently used for CNC machining, can extend the parametric relationship of design features directly to the CNC SM or AM machine toolpath. This parametric relationship allows changes made to the design model to automatically regenerate the laser path and control sequence sent directly to the machine. In comparison, when using an STL based file system, such as with most PBF-L systems, any changes to the design may require a redesign of the support structures and redevelopment of the laser path.

DED utilizing a closed loop real-time recovery, filtration, and reuse delivery system offers the potential to use a smaller total volume of powder thus decreasing the volume of reuse powder and allowing much higher powder to part volume ratios. Critical applications, such as outer space hardware or nuclear power systems, offering no opportunity for in service refurbishment, may specify the use of virgin powder with no opportunity for reuse.

Base features such as a plate or pipe may become an integral part of the final part. Repair of existing or substrate parts by adding a cladding layer or an additional

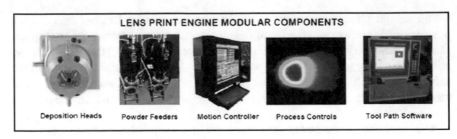

Fig. 8.17 LENS Print Engine Components[17]

part feature will enable remanufacturing or repurposing of existing parts and components. Multi-axis articulation without the confines of a powder bed will allow 3D scanning to determine part condition and orientation of repair parts, and to build upon the substrate the desired new feature or geometry.

High precision parts requiring post-processing to achieve final dimensions such as by CNC machining may not require the additional accuracy provided by PBF-L and could benefit from increased deposition rates of DED-L. If critical surfaces and dimensions must be machined anyway, the original accuracy of the as-deposited feature is less important. This is especially true for EB wire feed or arc based AM, where the benefits of high-volume deposition rates and material cost saving offset the need for high as-deposited accuracy.

The use of commercially available metal powder or weld wire filler material is a distinct advantage for DED-L versus PBF-L due to lower costs and the availability of a much wider range of certified powder or weld wire currently used in industry. Powder nozzle and delivery system clogging may still be an issue, but the overall powder requirements for DED are less stringent than for PBF-L or PBF-EB.[16]

The use of a base feature or build plate upon which to deposit may become an integral part of the final component, thus saving deposition time, post-process removal time, and material. In hybrid applications the base feature may also benefit from use of an automated stock feeding system further speeding production throughput.

Repair, remanufacturing, refurbishment, or enhancement of existing components is made easier by DED-L than by powder bed systems as the surface preparation, measurement, repositioning, deposition, finishing and in-process inspection may all occur in one setup, in sequence, on one machine.

Modular DED system components as shown in Fig. 8.17 will allow the repurposing, refurbishment or upgrade of existing SM type CNC systems or production lines, to include AM capability Hybrid AM/SM systems may be purpose built to accommodate specific fabrication tasks at a relatively low cost when compared to large general purpose systems.

---

[16]Personal communication with Richard Grylls, Optomec, (January 15, 2015).

[17]Photo courtesy of Optomec (reproduced with permission); LENS is a trademark of Sandia National Labs.

Hybrid systems incorporating AM with SM such as milling or turning, can reduce processing and setup time and be well suited for small lot size and small parts. In addition, DED can be used for dissimilar materials where cladding accuracy not as important, to repair worn or damaged components, to be applied to highly contoured surfaces or for difficult to process materials such as hard coatings.

DED-L systems and service provider Web pages offer system specifications, material data, and industrial application examples. Links to selected companies are provided for additional information in the AM Machine and Service Resource Links section at the end of this book.

8.2.2   Limitations of DED-L

As with PBF systems, the process is complex with many degrees of freedom or numbers of control parameters. All possible interactions of these control parameters both in a linear, nonlinear or chaotic manner is, quite honestly, mind boggling. These large numbers of control parameters can be an advantage or disadvantage. A better understanding of the process and control of these parameters can lock down or limit the degrees of freedom resulting in a more repeatable process. Figure 8.18 lists some disadvantages of DED-L.

The complexity of the process may be greater than that for powder bed methods primarily due to the software. Laser motion for traversing planar layers is inherently less complex than 3–5 axis simultaneous motion. But powder spreading and powder feed for DED is critical in either case. Laser powder interactions are in each

Fig. 8.18 Potential disadvantages of laser directed energy deposition systems

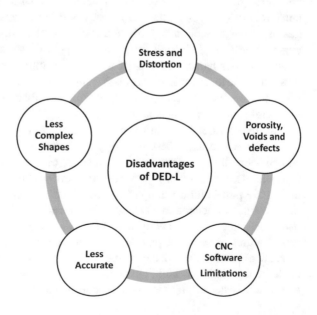

case complex. Hybrid machines such as those incorporating multi-tool turrets, stock material feed systems and those incorporating robotics significantly increase motion system complexity. Path planning for complex shapes puts the DED processes at even more of a disadvantage versus powder bed methods, as the degrees of process freedom and number of process variables and their interactions is huge. This complexity may have limited the industrial adoption of DED-L in comparison to PBF-L as applied to complex 3D parts.

The difficulty of powder recovery, recycle and chamber cleaning becomes more complex with larger glove boxes containing CNC motion control hardware.

The powder morphology requirements for DED may well be less stringent than with powder bed systems but powder flow characteristics and purity will still require high quality feed stock. Issues associated with powder recovery and reuse will be similar to PBF powder. As with PBF, economically viable sources for metal powder for all 3D laser metal printing processes must be established if these powder-based methods are to enter into widespread use.

DED-L may suffer from the same limitations as PBF-L, such as achieving the desired dimensional accuracy, surface finish and relatively slow build rates. In DED-L the larger molten pool, solidification and shrinkage stresses may result in higher levels of residual stress and greater part distortion.

As with laser cladding, wire feed delivery systems are also in use, although movement and articulation of a wire feeder may add complexity. In cases such as these, it may be better to articulate the part beneath a fixed head or articulate both the part and the head. The movement of massive assemblies, either the part or laser/wire feeder supply, requires large, rigid motion control systems limiting the speed to articulate, accelerate, or decelerate the mass of these assemblies.

Environment, safety and health issues associated with DED-L type systems may also need to consider the additional hazards of large build environments where entry by personnel is permitted. Powder and laser hazards, confined spaces, inert gas, mechanical motion hazards and lock out of equipment all require extensive engineering and administrative controls to provide for safe operation.

Added laser hazards may exist when using multi-axis systems. A fully articulated laser head capable of depositing material in off normal positions to a flat horizontal surface will require an enclosure not only capable of containing reflected laser light but also capable of withstanding direct beam impingement by a multi-kilowatt laser beam in the event of a motion system malfunction.

DED needs to use a base or support structure upon which to begin the deposit and buildup all subsequent features. In some cases this may be integral to the final components, but in other cases these support structures may need to be removed during the post build finishing operations. These support structures may need to be more robust to accommodate additional shrinkage forces and therefore may be harder to remove in comparison to PBF-L supports. See the discussion below of seed features or substrate parts for potential advantages.

Heat buildup within the part and within the built environment during a build could be a problem, as build operations can take hours and excess heat may be hard to extract. Heat buildup can damage equipment and create undesirable effects on grain growth, segregation of metallic impurities, formation of undesirable phases, defects, distortion and other metallurgical issues in the final part. Design complexity for DED may be limited in comparison to that attainable with PBF systems as the building of support structures may not be practical with DED for certain designs.

8.3 Additive Manufacturing with Electron Beams

Electron beam (EB) processing has the distinct advantages of high energy density (high beam quality, e.g., small spot size), high beam powers (multi-kilowatt) and is performed in a high purity vacuum, parts per billion (ppb) oxygen versus parts per million (ppm) levels as present in commercial welding grade argon. As with laser there are two basic types of EB based systems: PBF-EB and DED-EB. PBF-EB machines are currently produced by Arcam AB and referred to as the Electron Beam Melting (EBM) process. DED-EB machines are produced by Sciaky and are referred to as Electron Beam Additive Manufacturing (EBAM). A DED-EB process developed by NASA is referred to as Electron Beam Free Form Fabrication or EBF3.

PBF-EB using an electron beam is similar to PBF-L, as both start with a 3D model, create a deposition path by slicing an STL file and fuse powder material layer by layer, incrementing Z motion downward and recoating, and repeating the process until the desired shape is realized. As with PBF-L and DED-L, there are two methods to using electron beams, one fusing a bed of powder using a scanned beam source and the other articulating a mobile electron gun and wire feeder using CNC motion control.

The electron beam offers a number of distinct advantages and limitations when compared to both laser processing and arc based methods. The DED-EB higher purity vacuum environment offers a primary advantage by enabling the deposition of highly reactive materials and those susceptible to contamination by oxygen or other contaminants picked up during solidification and cooling. An additional advantage is related to the high beam powers achievable, large chamber sizes, and high deposition rates. EB systems have had the advantage of wall plug efficiency over lasers but the advent of higher efficiency diode and fiber lasers has narrowed this performance gap. Disadvantages are primarily related to equipment cost and complexity. First we will discuss powder bed methods.

8.3.1 Electron Beam Powder Bed Fusion Systems

An electron beam PBF process is shown in Fig. 8.19, a stationary electron beam gun may be attached and directed into a vacuum chamber containing a powder bed system with the beam electromagnetically deflected and scanned, in X–Y coordinates, on a flat build plane to trace out and fuse powder for each slice of the model. The ability to rapidly scan the electron beam using electromagnetic coils, as opposed

Fig. 8.19 The Arcam EBM process[18]

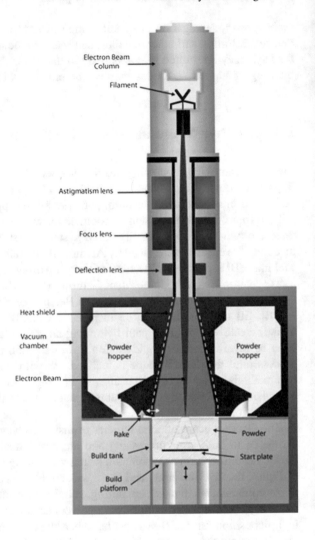

to moving mirrors in the case of laser PBF, allows faster build rates than similar laser scanned PBF systems. However, the process is limited to the deposition of electrically conductive materials. Advantages of PBF-EB are shown in Fig. 8.20.

Arcam AB[19] has commercialized the electron beam melting process for production of metal components using the EB melting of a bed of powder (Fig. 8.19). The technology offers freedom in design combined with attractive material properties and high productivity. Arcam emphasizes manufacturing in the orthopedic implant and aerospace industries. The EBM technology utilizes electron beam

[18]© Arcam, reproduced with permission.
[19]Arcam AB web page, http://www.arcam.com/, (accessed March 21, 2015).

Fig. 8.20 Advantages of the electron beam powder bed fusion process

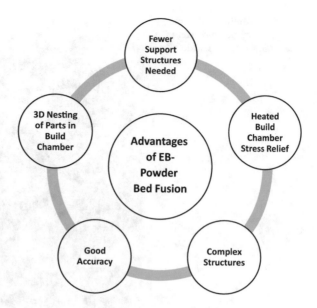

preheating of powders up to ~700 °C, maintaining temperatures that in effect stress relieve the parts during the build process. Camera based monitoring and a modular powder recovery system is provided. Arcam claims properties better than cast and comparable to wrought. The smallest focused beam spot size is on the order of 100 μm allowing the creation of fine details. Multiple melt pools can be maintained simultaneously due to the rapid electron scanning capability of up to 8000 m/s. Multiple parts can be produced during a build cycle offering high utilization of the build volume. Typical build dimensions are 350 × 380 mm. Helium gas is leaked into the chamber increasing working pressures to ~10^{-2} Pa and is used to reduce electrostatic charging of the powder particles and assist cool down after the build cycle. Arcam offers a validated supply chain for its powders primarily those of titanium and cobalt-chrome alloys and provides process parameters optimized for these powders.

The build chamber is typically heated to 680–720 °C and kept at the elevated temperature during the build. Preheating can vary for other materials such as aluminum (300 °C) or titanium aluminide (1100 °C). This serves as both preheat and post heat environment and helps to minimize shrinkage stresses and distortion upon cooling, residual stresses and the formation of non-equilibrium phases all of which can result in cracking of sensitive materials. Cooling of the build volume can take hours or tens of hours to cool. A defocus powder preheat pass is used to lightly sinter the powder and reduce the thermal gradients associated with the regions experiencing rapid heating and cooling surrounding the melt pool. The need for support structures and their post process removal may be reduced or avoided as powder adjacent to the part being built is lightly sintered during each layer, effectively serving as a support structure that is more easily removed and recycled

Fig. 8.21 Multiple instances
of a part may be fabricated in
one build cycle by stacking[20]

during post processing in comparison with the more rigid supports required by laser
powder bed systems. As with laser systems, the sieving of powder for reuse using
explosion proof vacuum cleaners and strict procedures is required.

The slow cooling cycle may be used to allow more time for grain growth and
relaxation of the microstructure both reducing locked in stress but also to reduce
distortion associated with the avoidance of localized shrinkage. Diffusion of interstitial contaminants or oxygen pickup during long cooling cycles may be a
problem for reactive materials. This may be exacerbated when a part is held at
elevated temperatures for long periods of time on the order of 8–10 h or more.

Arcam provides detailed material data sheets for titanium alloys and cobalt
chrome alloys. Mechanical property data is provided for EBM material and compared with cast and wrought properties for the various alloys. Post process heat
treatments and hot isotactic pressing can be used to improve fatigue performance.
Claimed advantages to this process include fast build speeds and the ability to stack
parts more easily within the build volume, as shown in Fig. 8.21.

Process limitations are some of those already described in the PBF-L discussions, such as the cost of the powder material and part size limitations due to
chamber size. Other limitations for PBF-EB (Fig. 8.22) include the time for the
build volume to cool from the high preheat and processing temperatures. Modular
build volumes can be removed and allowed to cool while a new build volume is
installed for the next build job. Fewer material options are available and the part

---

Fig. 8.22 Potential
disadvantages of PBF-EB

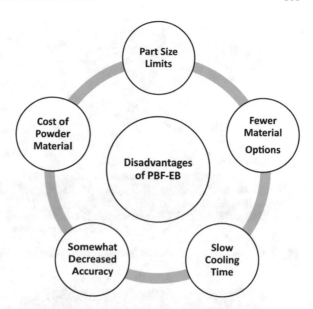

accuracy is slightly decreased due to somewhat larger powder diameter sizes.
Specialty powders with larger particle diameters (larger than with PBF-L) and the
electrical grounding of the build plate are required due to electrostatic charging and
repulsion of finer powder particles (often referred to as "smoke") disturbing the
powder layer. These larger powder sizes and the required focal conditions can
contribute to decrease in accuracy than is obtainable in certain laser based systems
using smaller diameter powders. The data sheets also specify a minimum particle
size of 45 μm for safe handling. PBF-EB powder sizes may be compared with other
PBF-L vendors stating powder sizes down to 10 μm. Video links of the EBM
process can be found here.[21] FDA approved PBF-EB implants are already on the
market demonstrating a clear path of adoption at the consumer level. A good Oak
Ridge National Lab (ORNL) YouTube Arcam video demo can be found here.[22]

8.4   Electron Beam-Directed Energy Deposition Systems

DED-EB systems integrate a mobile electron beam gun, CNC motion and a wire
feeder within a large high vacuum chamber allowing movement in X–Y or tilt
orientations to trace out and fuse a deposited bead of metal, one bead at a time,

---

[21]YouTube Arcam video link, https://www.youtube.com/watch?v=Wafws7FTwhc, (accessed
March 21, 2015).
[22]Direct Manufacturing: ARCAM mesh ball video, https://www.youtube.com/watch?v=
iegi6D5MKmk, (accessed March 21, 2015).

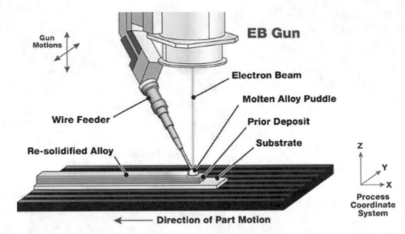

Fig. 8.23 Electron Beam Additive Manufacturing (EBAM™) Process[23]

Fig. 8.24 Electron Beam Additive Manufacturing (EBAM™) 110 System from Sciaky[24]

layer by layer. The Sciaky EBAM process is shown in Fig. 8.23. Using this approach, very large vacuum chamber build environments can be created (Fig. 8.24) allowing deposition of very large structures. High deposition rates are possible using a wide range of available wire alloys and sizes and choice of

---

[23]Photo courtesy of Sciaky, Inc., reproduced with permission.

[24]Photo courtesy of Sciaky, Inc., reproduced with permission.

Fig. 8.25 Advantages of electron beam directed energy deposition

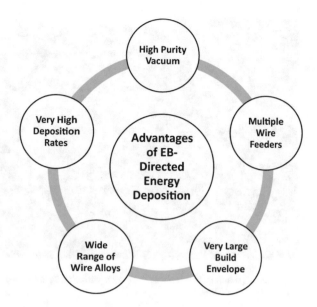

deposition parameters. Near-net shaped components display a distinctly stepped weld bead overlay shape that requires machining to create the final shape. Material selection is limited by the reliance on commercial sources of wire used by the process. One disadvantage is the slow cooling rate of the deposit within the vacuum environment and its potential effect on large grain growth and other metallurgical effects of the deposit. High degrees of distortion or residual stress may result when depositing large structures requiring post process heat treatment.

Sciaky's Electron Beam Additive Manufacturing (EBAM)[25] process is being marketed for use in the fabrication of large-scale, high-value metal parts using weld buildup to deposit shapes that can be made into prototypes or parts by subsequent post processing such as machining or forging. NASA has developed a similar process referred to as Electron Beam Free Form Fabrication (EBF3).[26]

Advantages include very large chamber sizes in comparison to the build volumes of powder bed type systems. Other advantages are listed in Fig. 8.25. Materials that are expensive, reactive, or of high melting points are attractive for use in the DED-EB process due to the capability of high beam power and the high purity vacuum environment. Parts and demonstration hardware have been produced in materials such as titanium, aluminum, tantalum, and Inconel. High melting temperature refractory metals, such as tantalum and reactive metals, such as titanium susceptible to very small levels of contamination by oxygen, have been successfully

[25]Sciaky EBAM Web link, http://www.sciaky.com/additive_manufacturing.html, (accessed March 21, 2015).

[26]NASA EBF3, Web link with demo video, http://www.nasa.gov/topics/technology/features/ebf3.html, (accessed March 21, 2015).

Fig. 8.26 A titanium hemisphere deposited with Sciaky's Electron Beam Additive Manufacturing (EBAM™) technology[27]

demonstrated as deposited by this process with vacuum levels in the $1.3 \; 10^{-5}$ mbar (10^{-5} torr) range. A view inside the vacuum chamber of the Sciaky EBAM machine shows the electron gun above a titanium hemisphere part and two wire feeders are shown in Fig. 8.26.

These machines can feature two wire feed systems capable of individually controlling each wire feeder allowing the creation of a graded deposit changing from one material to another. Deposition rates up to 6.8–18 kg per h (15–40 lbs./h) can be realized. The electron beam based process can provide advantages over current laser beam systems in beam power, power efficiency and deposition rate (Lachenburg 2011).

A Sciaky YouTube video link[28] provides a good view of the EBAM process in action. You can see the solid CAD/CAM model and deposition path simulation and the deposition as it proceeds. The large build chamber features a moveable electron beam gun and dual wire feeders. The deposited titanium metal remains bright with no signs of discoloration due to contamination as the chamber, wire, and material are kept very clean at all times. The video view during deposition shows the process proceeding smoothly without excessive vapor, spatters or ejected material. However, heat buildup in the part can be an issue as the vacuum environment limits convective cooling and may require long cooling times within the vacuum environment.

---

[27]Photo courtesy of Sciaky, Inc., reproduced with permission.

[28]Sciaky YouTube video link, https://www.youtube.com/watch?v=A10XEZvkgbY, (accessed March 21, 2015).

The size of the deposited weld bead, layers, and step size produce a coarser resolution of build shape that requires post deposition machining, but given the alternative of finding a large block of titanium and hogging it out with a machining process, DED-EB is in many cases may offer a better solution. Using EB welding to create a large thick section weldment from many piece parts, to form a structure then to be machined, may also have its drawbacks due to weld defect morphology and inspection limitations, making the DED-EB potentially attractive. DED-EB machines may be the largest and most expensive 3D printers currently available and undoubtedly a unique capability. A very wide range of metals may be deposited, but feed must be available in wire form. Deposition rates are quoted up to 4100 cubic centimeters (250 cubic inches) per hour or up to 18 kg/h (40 lbs./h) for titanium or tantalum.

Wire fed weld pools feature wire being consistently fed into the leading edge of the pool. Changes in direction require an articulation of the wire feed to optimize the consistency and control of the melt pool. As with all weld wire feed applications, wire feed irregularities due to coiling and straightening of the feed coil by the wire feeder can be an issue as is cleanliness or other dimensional variations. NASA is also working on real-time flaw detection and FEA predictive modeling of residual stress and is working with Virginia Tech to develop software that can help design and analyze lightweight panels, such as those fabricated with EBF3.[29]

NASA and other international space agencies are looking into the applications of 3D printing both in zero gravity space stations and on the lunar surface. Electron beam AM systems may also utilize the existing vacuum of the space environment while wire based systems would be easier to control than powders in a zero G environment. Electron beam systems are also much more energy efficient than similar laser based systems at the current level of these technologies. We'll talk more about space based applications later in the book.

Figure 8.27 lists some disadvantages of the DED-EB process. In addition, the difficulty of controlling the large melt pool can limit the deposition to the flat position. This may also adversely affect the resolution of smaller structural features, limiting deposition to bulk regions and straight walls and requiring the avoidance of overhangs. However, the possibility exists for the process to be stopped and support plates of run-off tabs to be added adjacent to the current layers of deposition. Wire feed systems for large part deposition require large continuous spools of material with the added complexity of large wire feed mechanisms. Variations in wire spooling and diameter can affect the accuracy of the wire feed which may wander in position during the feed process. Large, massive base plates or base features are required to help control the effects of shrinkage or distortion. Typical

---

[29]MSC Software web page link to NASA study, Subsonic and Supersonic Fixed Wing Projects—Virginia Tech and NASA, http://www.mscsoftware.com/academic-case-studies/subsonic-and-supersonic-fixed-wing-projects-virginia-tech-and-nasa, (accessed March 21, 2015).

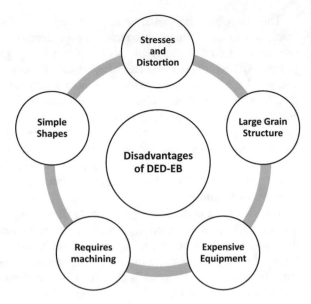

microstructures, fully evolved from the melt, display large grain size due to the high
heat input and slow cooling rates.

8.5 3D Metal Printing with Arc Welding Systems

Arc systems offer an affordable technology to achieve solid, fully fused near-net
shape metal objects. Arc and plasma arc (PA) based DED systems (DED-PA) does
not match the precision, accuracy or surface of PBF-EB or PBF-L but are able to
provide large near-net-shaped parts at a fraction of the cost. High end systems using
robotic arms or CNC gantry systems are able to achieve deposition rates and
accuracies of electron beam wire feed systems and are best suited to materials not
requiring the very high purity vacuum environment of DED-EB. Given the
decreasing cost of robotic systems, arc welding robotic 3D printers may one day
find a place within a metal fabrication shop near you in the near future. Some pros
and cons of DED-arc systems are shown in Fig. 8.28.

Cranfield University has developed such a DED-arc process, referred to as
Wire + Arc Additive Manufacturing (WAAM).[30] In 1994–99 Cranfield University
developed the process of Shaped Metal Deposition (SMD) for Rolls Royce utilizing

---

[30]Cranfield University, Revolutionary 3D metal production process developed at Cranfield,
December 16, 2013, http://www.cranfield.ac.uk/about/media-centre/news-archive/news-2013/
revolutionary-3d-metal-production-process-developed-at-cranfield.html, (accessed April 27,
2015).

Fig. 8.28 Pros and cons of directed energy deposition arc based systems

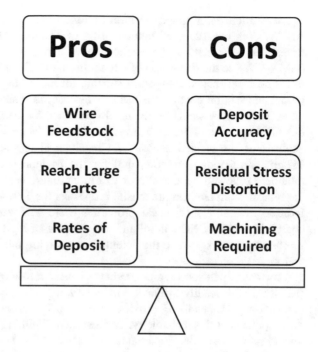

high deposition rate, high-quality metal additive manufacture using wire + arc technology",[31] accessing various arc based processes and materials for engine cases, with the primary objective of depositing large titanium alloy components. The benefits of using weld wire based AM were high deposition rates of kg/h, high material efficiency, with no defects and low part costs. Detriments to the process included the restrictions to depositing shapes of low to medium complexity, extensive distortion and large grain growth due to slow cooling rates. A degree of grain refinement was realized through changes to process conditions such as travel speed and the addition of a boron coating to the weld wire to nucleate grain growth and act as a grain refiner. In-process mechanical rolling of each deposited path was shown to induce cold work, grain refinement and recrystallization during reheating of subsequent deposition paths. As-deposited Ti-6-4 alloy material showed decreased strengths when compared to wrought materials. The addition of rolling the deposit between layers increases the yield and ultimate strengths of the deposit. Elongation properties varied for the deposit in the vertical and horizontal orientation to the build direction. In a case study presented for a specific component design the *Buy-to-Fly ratio* was reduced from 6.3 to 1.2 with a weight savings of 16%.

[31]Norsk Titanium News reference, Cranfield University, Colegrove, P., Williams, S., "High deposition rate high quality metal additive manufacture using wire + arc technology", http://www.norsktitanium.no/en/News/ ~ /media/NorskTitanium/Titanidum%20day%20presentations/Paul%20Colegrove%20Cranfield%20Additive%20manufacturing.ashx, (accessed April 27, 2015).

In an article describing the demonstration of a high level application, "Exploring Arc Welding for Additive Manufacturing of Titanium Parts" (Kapustka 2014) the researchers demonstrate the application of the gas metal arc/hot wire process (GMA-HW), to the deposition of a titanium alloy Ti-6-4 ELI (extra low interstitial) into a near-net-shaped component suitable for machining into a finished part. CAD model and motion control was used to deposit the shape onto a titanium substrate plate. This is very similar to the results obtained by the DED-EB process except it did not require a vacuum chamber or electron beam welding system. Chemical analysis of the welded deposit and mechanical test specimens were produced. Material properties were compared for the as-deposited condition, solution heat treatment followed by anneal heat treatment and anneal heat treatment along with the typical room temperature tensile properties for Ti-6-4 ELI castings. The results indicate favorable mechanical properties were achieved. This demonstration of combining CAD, CNC control and arc welding of high-value material to create a machining blank speaks to the potential of applying this process to other materials and more complex shapes.

Low cost, open-source, arc based systems are mashing plastic 3D printer motion (RepRap) with readily available GMA welding systems to realized shaped metal deposition.[32] Appendix E provides additional information regarding the recent developments in this technology. For a student, building a system like this is great way to start to learn the fundamentals of AM and DED systems, learn about system integration and heat sources. They you will also gain practical insight into metallurgical effects such as shrinkage distortion, part accuracy, and parameter selection. For anything but small objects, the mass of the object will adversely affect the ability of the modest RepRap type motion system to accurately move and articulate the object (or torch) during the build. Commercially available wire feeders such as a GTAW wire feeder upgrade, combined with a micro GTA torch and RepRap motion may be just the ticket for a do-it-yourself project for an entry level arc based AM capability. The US government program America Makes, mentioned later in the book, is helping to continue this R&D at Michigan Technological University, MTU.

Advantages of GMA-DED and DED-PA

Computer modeling and 2 ½ D slicing and path planning for motion control are well established for 3D printing of plastics and is, in the near term, directly transferrable for open loop motion control of arc based systems. The motion systems required for arc welding are cost effective enough for production level applications. Commercial level GMA weld system controls are readily available means to provide both arc control and feeding of filler material.

As demonstrated with DED-EB, machining blanks for large objects may be deposited without the need for a wide variety of commercial shape materials (e.g.,

---

[32]Link to and Open Source MTU 3-D metal printer combining RepRap and GMAW, https://www.academia.edu/5327317/A_Low-Cost_Open-Source_Metal_3-D_Printer', (accessed March 21, 2015).

plates, sheets, angle, I-beam, pipe) and the machines needed to process those materials such as shears, brakes and cutting tables. The prospect of reduced scrap may also factor into the utility of the process.

Disadvantages of GMA-DED and PA-DED

Wider spread usage of GMAW-DED system may also identify the need to harden or shield the motion hardware from the built environment from heat, weld spatter, or smoke particles. Heat generated by the process may damage joints and precision surfaces, requiring additional heat shielding.

GMA welding relies on wire feed and liquid metal droplet or spray transfer across the welding arc. This can result in weld spatter and a greater degree of particle and fume generation than arc systems such as GTA or PAW, where no metal is transferred across the arc. Proper inert gas shielding allows robots to deposit weld metal in bright clean beads at very high speeds, depending on the metal being welded.

Semi-automated wire feed and constant voltage power supplies control many of the process variables such as arc length and filler control, but the process often starts with excessive weld buildup and creates a lower penetrating more highly profiled, rounded weld bead. Fewer control options are available at the termination of the weld bead. Heat buildup in the fabrication of smaller parts or small part features will be an issue. Start/stop control issues (such as waiting for a part to cool) and providing protection from oxidation and atmospheric contamination while the part is being built. Part removal from the build plate may require sawing, milling. or larger machine tool capabilities unless it becomes integral to the final component. Developers of the technology will undoubtedly optimize heat treatment and HIP schedules to provide stress relief, more uniform properties given the range of microstructures typical to the base plates and weld deposits.

In an example of DED-GTA, Materials and Electrochemical Research Corporation[33] is offering rapid manufacturing of near net shape metal and alloys. Plasma transferred arc is a cost effective, less complex alternative to laser for solid freeform fabrication AM parts. The process features higher deposition rates, lower operation costs, higher efficiency and the ability to mix alloy powders and wires to achieve engineered functionally graded materials and surface layers and to include refractory alloys. Mechanical properties for Ti-6Al-4 V and Aermet™ 100 steel are reported to compare favorably to commercial grades of these materials. Various current and target applications and metallurgical data are provided.

Norsk Titanium[34] has developed a proprietary robotic based plasma arc Rapid Plasma Deposition™ (RPD™) process (Fig. 8.29a). This DED-PA AM process uses a patented torch design and control system to melt a titanium wire feed in

---

[33]Materials and Electrochemical Research Corporation, http://www.mercorp.com/index.htm, (accessed March 21, 2015).

[34]Norsk Titanium web page describing the RPD™ process, http://www.norsktitanium.com/, (accessed May 14, 2016).

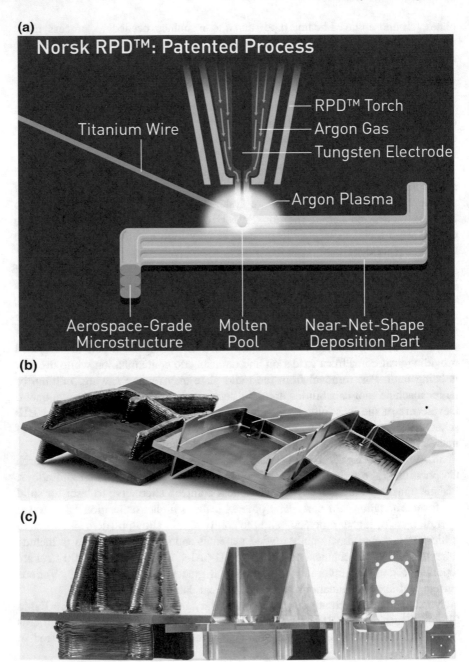

Fig. 8.29 a Norsk RPD™ process.[35] **b** and **c** As-deposited, partially machined and finished titanium parts[36]

argon, to build up layer by layer, a near net shape which is then post processed and machined into an aero-space grade titanium structure. Norsk claims production costs can be reduced by 50–70% when compared to legacy forging and billet manufacturing methods by significantly reducing material waste and the required machining energy. In addition the time to market can be reduced substantially. Target markets include aerospace, defense, energy, auto sport and marine markets. Figure 8.29b and c show an as-deposited shape, partially machined shape and finished component (left to right). Potential savings include reducing the lead time for replacement parts from months to weeks. Life cycle costs when switching from aluminum to titanium will allow aircraft to fly further, carry more passengers and cargo. Titanium offers greater compatibility when interfaced to carbon fiber composites than other competing materials. The buy-to-fly ratio can be significantly improved using the RPD technology when compared to historical methods. A single system is reported to be capable of depositing up to 20 tons of product a year with rapid changing over times and lot sizes of one.

Alternatively an AM deposited shape could potentially serve as a forging blank to be post processed using closed die forging within existing tooling. In certain cases where a stock of legacy forging blanks no longer exist, arc-weld-deposited near-net shapes may serve as blanks to be forged in a die rather than relying on the remanufacture of cast blanks that would require a mold and casting operation. This type of processing may serve to create components for the repair legacy systems where original component manufacturers no longer inventory forging blank stock.

Arconic,[37] split from Alcoa, features a wide array of material science, engineering and advanced manufacturing technology ranging from the production of aluminum and titanium powders and wire suitable for AM to through the development and applications of additive manufacturing, forging, casting machining heat treating and HIP technologies in support of aerospace, defense and space, energy, industry and transportation. Their Ampliforge™ developed by Alcoa,[38] combines AM processing with forging to produce forged properties in an AM deposited components.

8.6 Other AM Metal Technology

Figure 8.30 shows another group of additive manufacturing technologies that begin with a computer model and end up with a metal part.

Fig. 8.30 Other AM metal technology

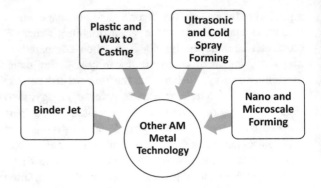

8.6.1 Binder Jet Technology

Binder Jet technology uses a powder coated with a binder, fusing the coated powder in a powder bed type system to form a green part. The binder is then driven off in a baking process and the part is infiltrated with a lower melting point liquid metal to create a metal powder/metal matrix composite part.

ExOne uses jetted-binder-on-powder technology which prints binder, layer by layer, onto a bed of powder. The powder box is then baked in an oven to cure the binder and the unbound powder removed to reveal the green part, which is then placed into a support medium, sintered within a furnace and infiltrated with a lower melting point material such as bronze to achieve densities up to 95%. A slow cooling annealing step is performed prior to final finishing. The process is being used for the bonding of stainless steel, iron, and tungsten. Process development is going as applied to other metals. Direct binder printing of sand cast molds are also being demonstrated for castings. A cast component is shown in the case study[39] of Fig. 8.31 where the Naval Undersea Warfare Center (NUWC)—Keyport needed replacement parts for an Ohio class submarine. A small lot of four vacuum cone castings of leaded red brass were required, size 11 × 5 × 10 inches. The OEM of these castings quoted \$29,562 each with a 51 week delivery time. ExOne using their Sand Printing Method produced and delivered the castings in 8 weeks for a cost of \$18,200 each using the reversed engineered CAD models supplied by NUWC. ExOne's web site has a good video demonstrating the process.[40] Binder jet technology may also be used with HIP processing to achieve near full density without the need for infiltration by other metals such as bronze providing additional dimensional allowance are made within the design to accommodate HIP process consolidation. Figure 8.32a shows a large titanium part produced using ExOne

---

[39]ExOne Case Study ©2014 The ExOne Company, www.exone.com, (accessed March 21, 2015).

[40]ExOne Web site, http://exone.com/en/materialization/what-is-digital-part-materialization/metal, (accessed March 21, 2015).

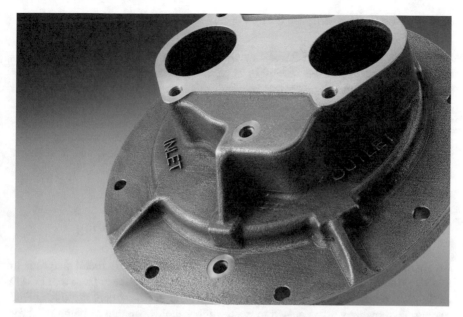

Fig. 8.31 A cast component produced using the ExOne Sand Printing Method[41]

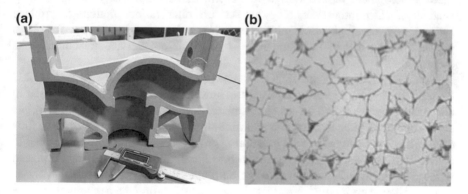

Fig. 8.32 **a** A large titanium part produced using ExOne binder jetting and HIP processing to achieve full density.[42] **b** The fully dense microstructure showing the light primary alpha regions and dark regions of intergranular beta phase[43]

binder jet technology and HIP processing to achieve a fully dense part. Figure 8.32b shows the fully dense fine microstructure of the titanium part where light regions are primary alpha phase and dark regions are beta phase.

---

[41]Courtesy of ExOne, reproduced with permission.

[42]Courtesy of Puris, reproduced with permission.

[43]Courtesy of Puris, reproduced with permission.

8.6.2 Plastic Tooling in Support of Metal Fabrication

Plastic tooling in support of fabricating thin metal parts by bending, stretching or hydroforming may be used in certain application and material dependent operations. Plastic jigs and fixtures may be appropriate for certain conventional metal processing applications and in some cases may replace fixtures historically made of metal. One vendor cites the ability to produce dissolvable plastic bending mandrels. Any of the plastic prototyping technologies may be applied depending on the requirements of the deposit properties such as strength, accuracy, speed, or other performance criteria.

8.6.3 Plastic and Wax Printing Combined with Casting

An alternative way to use 3D printing technology to realize fully metal shapes is to use 3D printing of plastics, polymers, or wax to produce a pattern from a 3D CAD model. The pattern may then be used to create a sand or investment mold for metal casting. Patterns may also be used to make rubber molds capable of being used for low melting point materials such as lead, tin or zinc. These patterns may be used to speed development of and design refinement of patterns. Many of the leading vendors of 3D printing technology or services offer various options to produce patterns or even single use molds. To understand more, survey the Web links provided at the end of the book in AM Machine and Service Resource Links section.

In a 3D Systems case study,[44] QuickCast patterns were used by Tech Cast LLC to speed the design of large complex pump impeller castings weighing up to 350 lb and measuring up to 30 inches in diameter. The rapid process allows multiple design iterations to be evaluated at one time significantly reducing the time to final design, thereby saving on prototype tooling costs and lead time when compared to conventional methods. Patterns are resin based and can be designed and fabricated to result in accurate dimension and low burn out ash.

In another case study[45] Voxeljet describes the casting of a complex bronze object using a lost plastic 3D printed model and the lost wax process. The final part, shown in Fig. 8.33 is 276 × 239 × 221 mm and weighs 10.5 kg.

---

[44]3DSystems, QuickCast case study, http://www.3dsystems.com/sites/www.3dsystems.com/files/tech_cast_case_study.pdf, (accessed August 13, 2015).

[45]Voxeljet case study "Knot", http://www.voxeljet.de/uploads/tx_sdreferences/pdf/plastic_model_knot_ENG_2012.pdf, (accessed August 13, 2015).

Fig. 8.33 A complex bronze object cast using a 3D printed plastic model and the lost wax casting process[46]

8.6.4 Ultrasonic Consolidation

Ultrasonic consolidation (UC) or Ultrasonic Additive Manufacturing (UAM) is a layer based additive/subtractive manufacturing process that begins with a solid computer model, slices layers of material and creates toolpaths for an ultrasonic tool to bond the layers of metal or other composite materials into a consolidated shape. Subsequent milling or machining operations are required to remove the part from the base plate and remove the bulk of unfused layers surrounding the solid region to realize the final part shape. Aluminum, copper, and titanium alloys have been successfully joined using consolidation, although not all metals may be used with this process. The process can join dissimilar metals and materials that may not be processed using conventional melting or AM processes relying on melting as this is a solid state bonding process producing little heating. Fabrisonic is one company featuring this technology.[47]

8.6.5 Cold Spray Technology

Cold spray is a process where metal powders are sprayed at high velocities to cold weld the powder onto an existing substrate surface. It may be used to build up shapes or rebuild surfaces. It can produce microstructures not achievable with

---

[46]Courtesy of Voxeljet AG, reproduced with permission.

[47]Fabrisonic Ultrasonic Additive Manufacturing, http://fabrisonic.com/ultrasonic-additive-manufacturing-overview/, (accessed April 10, 2015).

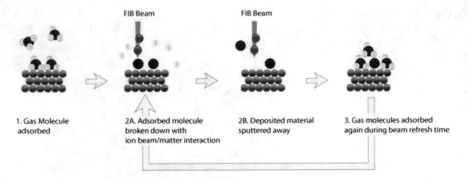

Fig. 8.34 Focused ion beam principal[51]

solidification based processing and has been referred to as "3D painting".[48] When coupled with subtractive methods such as CNC milling, accurate shapes may be formed. Cold Spray Consolidation has been demonstrated for use in coatings and simple shapes in a project funded by the European Space Agency.[49] The Commonwealth Scientific and Industrial Research Organization (CISRO) of Australia have developed a number of cold spray applications[50] such as titanium heat pipes and other free form components. In addition the technology has been demonstrated for a wide range of coating and repair applications.

8.6.6   Nano and Micro Scale Methods

Nano scale is a term, when used in the context of AM that can refer to shaped deposits or structures of dimensions below 1 μm, while microscale refers to deposits and structures deposited in the 1–100 μm range. As an example, the focused ion beam lithography process (Fig. 8.34), can create very small precise features and structures on the scale of nanometers (Fig. 8.35). Micro-electro-mechanical systems (MEMS) use semiconductor device techniques to fabricate machines in the micro scale, nano-scale devices are referred to as (NEMS). MEMS may be deposited in metals using sputtering, evaporating or electroplating. Although these processes are highly precise, they are only used for very small

---

[48]Additive Manufacturing article, May 8, 2014, http://additivemanufacturing.com/2014/04/08/ges-cold-spray-provides-a-new-way-to-repair-and-build-up-parts/, (accessed March 21, 2014).

[49]Enginerring.com article, http://www.engineering.com/3DPrinting/3DPrintingArticles/ArticleID/9419/Engineers-to-Fine-Tune-Cold-Spray-a-Next-Gen-3D-Printing-Technology-for-Astronauts.aspx, (accessed March 21, 2015).

[50]CISRO web page, http://www.csiro.au/en/Research/MF/Areas/Metals/Cold-Spray, (accessed January 31, 2016).

[51]Courtesy of Fibics, reproduced with permission.

Fig. 8.35 Focused ion beam nano-machining or deposition[52]

objects due to deposition speed limitations. Although build speeds are slow, large numbers of devices may be made at the same time.

As an example, Microfabrica makes high-volume production of micro scale metal parts in the micron to millimeter range with extreme precision as shown in Figs. 8.36 and 8.37. Their MICA Freeform process can fabricate precise holes, channels, ribs. Mechanisms can be formed in their assembled states, all in one shot. Engineering grade metals deposited include Valloy 120 (nickel/cobalt), palladium, rhodium, and copper. The Microfabrica Web site has a good video.[53] The process can fabricate micro devices with all moving parts fully assembled with features as small as 20 μ and with tolerances as close as ±2 μ.

8.7 Key Take Away Points

- The two main categories of AM metal systems are referred to as PBF systems and DED. PBF and DED systems are further differentiated by the heat source used, laser beams, electron beams and plasma or electric arcs.
- PBF systems feature greater accuracy but are constrained by the powder bed size. DED EB and arc based systems generally offer a larger build envelope and faster build rates but at the cost of resolution and accuracy often requiring 100% post process machining.

---

[52]Courtesy of Fibics, reproduced with permission.

[53]Microfabrica Web site, http://www.microfabrica.com/, (accessed March 21, 2015).

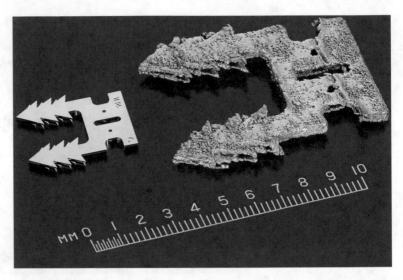

Fig. 8.36 A part made by Microfabrica Inc. compared to a part made using metal laser melting[54]

Fig. 8.37 Implantable medical electrodes made by Microfabrica Inc[55]

- Plasma and arc welding systems allow the deposition of large near net shaped parts and are most easily adapted for use with robotic motion.
- Other AM metal processes allow the deposition of very small components, while others use AM patterns or models to be used by conventional casting or

---

[54]Courtesy of Microfabrica Inc. (Microfabrica.com), reproduced with permission.

[55]Courtesy of Microfabrica Inc. (Microfabrica.com), reproduced with permission.

forging processes. These processes may offer lower cost alternative to direct metal PBF or DED processes, particularly for large components.

- An in-depth understanding of the advantages and disadvantages of these different AM metal systems, and the alloys they are best suited to process, is an important step in selecting the appropriate AM process for a particular design. Fig. 8.1 Additive manufacturing metal processes.

Chapter 9
Inspiration to 3D Design

Abstract A strong attraction of solid freeform fabrication is the removal of design constraints imposed by commercial metal shapes and conventional processes; opening a whole new world of design possibilities. Commercial metal shapes associated with sheet metal, tubing, angle iron, pipes and plates will not be going away anytime soon, but add the possibility of forming flowing organic surface shapes, complex passageways, and unique internal or external structures, and we can begin to rethink what is possible when working with metal. The same thinking can release us from shape constraints imposed by creating features using drilling, shearing, bending, milling and casting or pressing into molds. This chapter will describe the AM design process and compare it side by side with the design process for conventional metal processing. We will include a few examples of how one method is better than the other and which is right for you. We will discuss the advantages and disadvantages of selecting one material over another. In addition, we will highlight a few examples of hybrid applications, combining the best of 3D printing, conventional and subtractive processing and how 3D metal printing can define a new design space for thinking outside the box. This chapter will build upon the existing body of metal working design knowledge, complement it, transform it and take it to levels not possible in years past.

9.1 Inspired Design

A mechanical engineer will often start the design process with a list of requirements such as end use, part function, size, material, weight, strength, dimension accuracy, system compatibility, and cost. This process is needed for both large and small components but is largely driven by how important or critical the in-service function of the part is. Critical applications such as those in aerospace, automotive, or the medical field will place part quality, certified materials, processes and design ahead of factors such as cost or aesthetics. The value added by certification can often exceed the fabrication cost of a component, therefore if the part requires certification, the materials and process will as well. Conversely, jewelry is example

© Springer International Publishing AG 2017
J.O. Milewski, *Additive Manufacturing of Metals*, Springer Series
in Materials Science 258, DOI 10.1007/978-3-319-58205-4_9

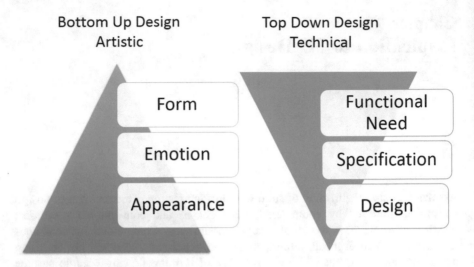

Fig. 9.1 A comparison of artistic verses technical design

of noncritical application where the primary design requirement of the artist may be aesthetic or emotional appeal.

As stated in "Emotional Design", (Norman 2004), "we perceive attractive things really do work better and are used more" as "emotional appeal can often override functional or utilitarian requirements". His book describes design from two perspectives, top-down and bottom-up (Fig. 9.1). Top-down design begins by thinking, how should a design function, what is the meaning of the design, how should it behave? Top-down design is cognitive, questioning the effective use. Bottom up begins by how one perceives and feels about the design asking what the visceral response is. Bottom-up design is affective; how does it appear to you? Norman talks about designs being appropriate for a time, place, and audience.

An example of how AM could take this emotional design concept to the next level would be user-defined souvenirs. One can imagine mobile apps that allow the creation of virtual or "parametric kitsch" where the importance of the time or place of an event, such as a major sports victory, combined with an individual's presence, preference or team identity could be combined using color, size, material and personalization. The power of this personalized design freedom could be described using words like symbolic, evocative, sentimental, talisman; something that engages one's attention, reaffirms one's identity, tying their thoughts to time and place. A mobile on-demand app could recognize time and place and the mobile user could input personalized information and press the send button to have the memento show up at his door the next day (Fig. 9.2). While not all of us want our face on a bobble head, there are other individualized experiences that enable the same level of engagement.

Another form of inspired design requiring personal engagement and assistance comes in a kit form. The Heathkit (from the Heath Company) or the Betty Crocker

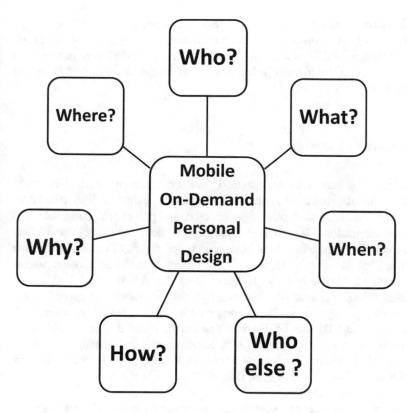

Fig. 9.2 Mobile on-demand personal design considerations

cake mix of the 1960s offered the techie or homemaker an option to participate in the process of creation, offering a guided hands-on challenge and the pride of accomplishment. The Altair computer was another example[1] of the attraction of hands-on creation. Personal 3D printer kits on the market today are offering the same level of active participation.

A balance often needs to be found between design complexity and fabrication capability. Realizing the full benefits of AM design can be daunting for both the experienced designer and a new designer given the part complexity achievable by AM. Both high end and entry level design tools will evolve to reduce the complexity of use and allow the maker to focus on the design itself rather than the workings of the design tools or manufacturing processes.

---

[1]As described in the Make: magazine article, Vol. 42, Dec 2014/Jan 2015.

GE refers to the concept of the Brilliant Factory in a press release,[2] developing a suite tools to integrate, software, data, advanced analytics, sensors and the cloud, support engineering collaborations and crowd sourcing design. This process can employ virtual testing and ultimately download final designs to intelligent machines for manufacture.

9.2 Elements of Design

What considerations need to be made in designing a component for AM fabrication versus conventional processes? Here we describe the design to fabrication flow beginning with the initial concept, design, material, and process selection, followed by prototype fabrication. Understanding the process flow for conventional metal processing will help you understand how solid metal freeform design is in some ways similar, but in other ways radically different. Sometimes a better understanding of what is "in the box" helps one to think "outside the box".

As stated above, elements of artistic design take a bottom-up approach and appeal to artistic perception and the senses, evoking an emotional response, both within the artist and the beholder. Elements of artistic metal design include inspiration captured in the material, shape, form, color, texture, and perhaps movement. Envision a fine gold jewelry pendant, or a massive Picasso metal sculpture in a city plaza, or a kinetic piece of junkyard garden art; the artist is always both limited and unlimited in her or his creation.

The elements of engineering design take a top-down approach and begin by considering what is needed; what is the effective use? An iterative, circular, and repetitive process imagines a solution, asking has this been done before and how it can be done now? A path is chosen among ideas to plan and consider what materials are possible, what processes exist to work these materials and what resources are available. The answers are folded into a design that balances and optimizes requirements with resources and constraints. Often the design is then reviewed by the stakeholders and either sent back to the drawing board or fabricated as a prototype and tested.

Designers with experience using conventional metal processes often have a family of component types and performance history to draw upon. Engineering text books or Web sites are full of part performance data, material performance, well-specified processes and procedures. These resources provide the designer with a well-known starting point that simply requires modifications to existing standard practices. It is not uncommon to modify an existing design to accommodate a new product line or material, to be manufactured in the same facility, using the same machines and skills.

---

[2]GE press release SAN FRANCISCO—Sept. 29, 2015, https://www.ge.com/digital/press-releases/ GE-Launches-Brilliant-Manufacturing-Suite, (accessed May 14, 2016).

In comparison, AM has little of this upfront knowledge and experience, although more is being generated every day. Corporate R&D labs using AM are making parts and subjecting them to analysis at an increasing rate to answer questions related to material and part performance. Successes in prototype part performance are published in featured articles on a daily basis although notable public failures have yet to emerge at this stage of development and acceptance. AM metal parts may never compete with certain mass-produced components using existing commercial methods, but it is hard to argue against the cost savings realized by cutting a product development cycle from months to weeks, even if material costs are inordinately high.

9.2.1   Material Selection

As mentioned earlier, commercial metal stock can range from sheet, pipe, channels, angel iron, I-beams, etc. The properties, chemical composition ranges and weldability are well understood for these forms. It is not perfect, but for the most part it works. The same goes for casting alloys and commercial powder used by conventional powder metallurgy processes. Metal in commercial shapes is a known place to start for conventional processing. Thanks to industrial standards you have a pretty good idea you know what you are getting. There is a huge body of knowledge developed over the past century in fabrication using these commercial shapes and alloys. For critical applications, material heats, lot numbers and chemical analysis is provided and can be preserved to track the pedigree of the material, offer traceability and allow comparisons between vendors. As an example, recommended practices for welding are well established for weldable metals which make setting up the process pretty straightforward, as long as one is willing to take the time to read up. However, one still needs to begin with a proper design and actually know how to fabricate the part.

Filler wire and electrode wire has evolved for welding applications over the past century and have been optimized for joining commercial base metals and for use with specific weld processes. AM applications have generally used spooled bare wire, either as an electrode (as in GMAW) or introduced as a filler into the molten pool (as in GTAW or PAW). Commercial weld filler metal is generally selected based upon the base materials being welded. If a weldment is entirely composed of weld metal, as in AM, then those properties quoted in the data sheets are a starting point only. The chemical composition of these wires have been optimized to accommodate cooling rate, loss of minor alloying constituents during transfer across the arc, or to accommodate the loss of strength in the weld and heat affected regions when compared to the base metal. Accommodation of weld wire chemistry to account for microstructural variations between the base metal and deposit do not exist for AM as the entire part is made up of the deposit. Wire spools and wire feeders can be small enough to be positioned on a handheld gun or remote to an automated system in large quantities for production applications. As with other commercial metal shapes, the pedigree of these filler wire materials may be preserved for qualification records as required by critical applications.

What wire fillers are available? The best place to start are the welding consumable data sheets provided by the vendors on their Web sites and within their product literature by following the AM Machine and service resource links at the end of this book. Weld consumables vendors provide material properties and nominal chemical composition, as well as design and process considerations such as process torch types, parameter guidelines and post-processing requirements such as heat treatment. Prices can range significantly depending on the alloy and specification. Specialty wire producers can supply wire for materials not typically welded, but these may not be suitable for AM applications and would have to be evaluated on a case-by-case basis.

Certain high-strength alloys specify component preheat, inter-pass temperature and post heat treatment to allow controlled cooling and time intervals to allow hydrogen diffusion and release to avoid cracking after welding. How AM will reliably achieve these conditions given differences in the these processes remains an open question and this may well dictate limitations of using certain weld filler metal materials for AM.

As mentioned earlier, specialty metal powders have been optimized for use in AM to assure flow and other characteristics such as bulk density. High purity, highly spherical powders with controlled diameter ranges are offered and in some cases specified by vendors to assure process repeatability associated with proprietary build parameters. These powders command a premium price due to the cost of production and demand due to rapid adoption and growth of the AM industry. To date AM vendor-certified powders are limited to a few of the most common engineering alloys of steel, titanium, nickel super alloys, and aluminum. Powder sizes for PBF-L typically range in the 20–35 µm range, while powders for laser DED-L range in the 35–100 µm range and powder for PBF-EB are within the 80–100 µm range. Certain vendors use smaller or larger powder sizes depending on the machine and applications. Table 9.1 lists the most common types of metal alloys offered by AM machine vendors and the applications in which they are often used.

What powders are available? The best places to start are the metal powder data sheets provided on the Web sites of the AM machine vendors and within their product literature. They can provide material properties, nominal chemical composition, as well as design considerations such as minimal material wall thickness and post-processing requirements such as heat treatment. Prices and availability are available upon request. Prices can range significantly ranging from $100/kg to ~ 1000/kg depending on the alloy and specification. The next place to look is on the metal powder producer's and vendor's Web sites. Many of these producers are now adding links to advance manufacturing or the specialty AM process powders they currently offer. Specialty powder producers can supply powders not typically used in AM but these may not be suitable for your application and would have to be evaluated on a case by case basis.

Uses of noncertified AM powders add additional risk to the development and repeatability of the AM process as these processes can be highly sensitive to

Applications for the Common Additive Manufacturing Metals

Alloy Type	Aluminum	Maraging Steel	Stainless Steel	Titanium	Cobalt Chrome	Nickel Super Alloys	Precious Metals
Aerospace	X		X	X	X	X	
Medical			X	X	X		X
Energy, Oil, Gas			X				
Automotive	X		X	X			
Marine Environment			X	X		X	
Machinability Weldability	X		X	X		X	
Corrosion Resistance			X	X	X	X	
High Temperature				X		X	
Tools and Molds		X	X				
Consumer Products	X		X				X

powder purity and other powder characteristics. Mixing of AM powders between different lots and heats may void the material pedigree associated with a certified process. With the current limited range of AM powder process metals, it may make sense to use recycled powder or powder that has lost its pedigree for the prototyping stage of development and upgrade the material when entering production qualifications stages of process development. These will be tough choices for engineers and business owners attempting to remain both profitable and agile while trying to keep up with the competition.

Industrial gases used in AM come in various purities and are typically provided in pressure vessels such as compressed gas cylinders. High purity (~ 2 ppm O_2, 10 ppm other) or welding grade (~ 5 ppm O_2, 40 ppm other) inert gases such as argon may be specified. Large production operations may require large trailers to be positioned external to the building with gas being delivered by plumbing. Depending on the material form (powder versus wire) and the AM process, nitrogen generators may be integrated into the AM machine to supply process gas. Partial vacuum processing chambers may be an alternative for materials that require high levels of atmospheric purity while reducing the costs of high purity inert gas usage.

Direct recycle and reuse material is being proposed as a sustainable feature of AM processing. While the use of recycled material has been demonstrated for plastics, direct metal recycle other than from AM powder reuse remains as a topic of study for the supply of metal for AM. As mentioned earlier, the secondary metals market currently exists to blend various grades of salvage metal into the primary metals stream, but this is not the general practice of AM powder vendors. Analysis of secondary powder material usage in AM is ongoing in order to determine the pickup of contaminants will reach unacceptable levels or other changes to the powder morphology when reused multiple times.[3]

Repair, remanufacture or modification of components that have been in service is an important role for AM processing. Renewal of metal components subject to wear, breakage, corrosion, or a host of service conditions is motivated by the high cost of replacement or the potential for improved performance and service life extension. The use of weld cladding to renew everything from train wheels to jet turbine blades and marine shafts is common practice as the utility has been widely demonstrated. AM offers the potential to extend that repair capability to make even more complex repairs to parts. Preserving information regarding a component, such as the CAD design and the service life conditions it operated under, may better allow repair, renewal, and remanufacture using AM. Cleaning and preparation procedures developed specifically for AM may further enhance the attractiveness of this type of repair. Access to original CAD model part definition, combined with 3D part scanning and automated decision-making, is needed to characterize, plan, and complete the repair using autonomous AM. The ability to access the original

[3]LPW Technologies reference to presentation at Rapid 2014 and software offered to maintain traceability, document powder aging and highlight contamination prior to powder use, http://www.lpwtechnology.com/lpw-technology-presents-new-research-recycling-additive-manufacturing-powders-rapid-2014/, (accessed March 21, 2015).

digital design definition, throughout the part's service life, will speed development and adoption of remanufacture methods for the repair of failed or broken parts.

The use of designs that utilize multiple and engineered materials has been common in applications such as weld or laser cladding for applications requiring hard facing or corrosion resistance. AM promises to take this to another level opening up the design space to the possibility of using multiple materials, graded materials or hybrid materials to build complex shapes. Service providers currently offer multi-material, cladding or graded material repair.[4] DED-L systems using multiple wire or powder feeders provide the greatest utility and material selection for repair, due to part access. AM use of composite materials and high-performance plastics can in some cases provide a fully functional part fixture using 3D plastic printing technology for parts that were historically made using metals. This may be an attractive alternative for limited production fixtures or small lot sizes

9.2.2   Process Selection

Once you have the functional requirements of a design and a good idea of the material requirements, you need to consider candidate processes. Earlier in the book we described the main AM metal processing systems, their pros and cons and how they are best applied. This knowledge combined with your design concept and material will assist in selecting which of these process options are best for you.

Part size, material, application, and service requirements all come into play when selecting a process, as do available AM services and their resources. What service providers are available to meet your needs? The plastics and metal prototyping industry has evolved to supply immediate online quotes for on-the-spot pricing and delivery. The material prices and delivery options will become more attractive for metals as more machines and service providers become available. Providers are still on the learning curve for how to price and fabricate parts across the range of materials and shapes currently offered. Established service providers can better estimate the contingency costs associated with failed build attempts or the need for multiple trial builds to optimize part orientation for printing. The best place to find an AM printing service provider is a Web search. We list a few in the AM Machine and Service Resources links toward the end of this book without giving our specific endorsement. Be sure to check the Web as new and more cost-effective options will continue to become available in this rapidly changing field.

AM machine vendors often have preferred service providers or in some cases offer the service themselves. Machine vendors with relationships with service providers often offer the most highly evolved process streams. The details of the process are often obscured from the buyer and considered proprietary to the vendors

---

[4]DM3D Technology LLC provides a wide range of cladding repair services using DMD® technology, http://www.dm3dtech.com/, (accessed March 21, 2015).

and machine builders as the value of intellectual property is very high at this early stage of process development in this emergent marketplace. Protecting intellectual property (IP) associated with the processes and materials is a never-ending battle with customers who want transparency into the working of the process and companies trying to develop and protect their own technology.

For small to mid-size business owners, conventional CNC processing often wins out based on familiarity, material options, and local resources. But with the advent of lower cost 3D CAD software, small to mid-size businesses often do start with a model and may either submit the model for an AM fabrication quote or turn it into an engineering drawing for a quote at the local CNC machine shop, or both.

Adoption of AM processing will increase as more users ascend the learning curve by jumping through the AM hoops enough times to gain confidence and weigh the longer term benefits with the near term investment. Many small businesses have incurred losses by an underestimation of the upfront cost and the learning curve to establish a new capability or new vendor stream. It can be hard for highly capitalized businesses to remain agile in a rapidly changing manufacturing world. Unless you are a major corporation, or AM service provider, your best near term decision may be to rely strictly on the resources of AM service providers rather than buy a machine that may be obsolete in 5 years. Appendix G provides a scorecard to help you access your AM related skills and your readiness to adopt AM technology.

Process and vendor selection may be affected by other factors as well. If you already happen to have a heat treatment furnace, a HIP, or are capable of post-processing or finishing as-built AM parts in house, this may also work to your advantage and alter your selection process. As discussed later in the book, your company knowledge and ability to do the upfront AM design work, with the proper tools and skills, may also be a consideration. How much post-processing must be done? Who will do it? Are there any material, design, or process selection considerations affected by the need for post-processing? The conceptual design, materials, and process selection may be iterated, modified, and down selected until you are ready to begin a formal AM design, provided your design tools, knowledge, and skills are up to the task.

9.3 Solid Freeform Design

Solid freeform design is a catch-all term that evolved from the desire to be removed from conventional design constraints yet be provided with functional bulk materials. The freedom of AM design from conventional form, has taken us far beyond the constraints of commercial shapes, molds, dies, and tool geometry. In addition, AM-deposited materials exhibiting porous or repetitive engineered substructures, such as lattice or honeycombs, take us far outside of conventional thinking when describing uniform bulk material. As you will see below, AM process constraints must restrict the final form of the design to what is achievable by the AM process being used.

Fig. 9.3 Hip implant
featuring complex porous
structure for bone in-growth[6]

AM can deposit porous structures that offer benefits for filtration or osteopathic bone ingrowth as shown in Fig. 9.3 weight reduction, or a host of other functions. Additional benefits may be realized by designing structures to provide directionally dependent mechanical or thermal properties. In addition, part design may include functional performance features such as those that provide for energy absorption, gas or fluid flow control, or acoustic wave propagation. How do you get your head around all the designs and possible applications for complex internal passages, lattice structures[5] (Fig. 9.4), honeycomb structures and engineered surfaces? What is possible using AM? What are the limitations? Do I have the tools, much less the know-how?

Design complexity is a double edged sword. While there may be great benefits, the difficulty of understanding, optimizing, and fabricating complex designs using AM can be a daunting task. Design complexity is not free if clever designs are difficult to build or finish or continue to fail in prototype testing. Optimizing the selection of literally hundreds of design variables may be beyond unaided human capabilities and at best a guess. It is inevitable that computer-based optimization algorithms for AM will be driven by the need for better AM processes. As described earlier in the book, we will undoubtedly see software emerge to provide a combination of multi-physics, multi-scale modeling, reduced order modeling and Big Data mapping to predict how best to design the part and process. Experience will help us to develop process maps to aid in parameter selection and design decision-making. Group sourced designs may also come into play, but until then, we learn by trial and error, some of us becoming more expert than others. As we develop the tools to assist us in optimizing these processes we will indeed see a paradigm shift in how we create 3D functional objects.

Design Tools

As discussed earlier in the book, 3D solid modeling software has been with us for some time now, optimized for use with conventional processing, taking us from

---

[5]EPSRC Web site, ALSAM Aluminum Lattice Structures via Additive Manufacturing, http://www.3dp-research.com/Complementary-Research/tsb-project-alsam-/13174, (accessed April 27, 2015).

[6]© Arcam, reproduced with permission.

Fig. 9.4 Nine gyroid cellular lattice structures 25 × 25 × 15 mm built on a base plate by the SLM process using stainless steel[7]

design through CAM with an industry focused primarily on conventional processing. Fully parametric model based engineering can carry a CAD model through to FEA analysis to CNC machining and inspection. Direct metal deposition and hybrid systems can make use of these conventional CAD/CAM methods, although some machine vendors are offering application specific software to allow improved ease of use or offer enhanced AM capabilities.

Model translation to STL file format has been the mainstay of powder bed systems with software developed to allow the fixing and editing of the STL file, design of support structures, perform slicing, and translation to machine specific instructions. Varying degrees of capability are provided with software tools ranging

[7]*Source* Chunze Yana, Liang Haoa, Ahmed Husseina, David Raymon, International Journal of Machine Tools and Manufacture Volume 62, November 2012, pp. 32–38. Reproduced with permission.

in price from free open source software to high-end engineering environments costing tens of thousands of dollars.

A place to start with the selection of AM design tools is to assess your current modeling capability. Most CAD packages now offer STL file translation to one degree or another. If you have plans to buy a 3D metal printer you may need to extend your modeling capability by buying more software to enable building your parts in house. STL file errors such as non-watertight designs, gaps, shared edges, inverted normal vectors, non-volumetric geometry and scaling issues will prevent the STL file from printing. These errors can crop up during design or translation to STL format and must be resolved before creation of the final control file. The creation of honeycomb, porous, or lattice structures may not be supported by your software and may require the purchase and use of additional third-party software. If you plan to utilize 3D metal service providers, you can simply pass along your STL model for modification and fabrication and live with the result. For many, this is a good place to start.

Creating an AM Design

By now you have your design concept, design tools, have chosen a material and a process and have considered the basic needs of post-processing. For this discussion we will describe an AM CAD design using a parametric, feature based approach. Not all CAD software supports this functionality, so ultimately you will have to work within the limitations of your CAD capability. As described earlier in the book, parametric models allow dimensions to be defined using variables instead of fixed values. This allows subsequent adjustment of the model shape by changing the values of the variables within limits and regenerating the model to render its new size and shape. Features may be defined in a parent–child relationship and as additive or subtractive volumes. As an example, a flange may be a parent feature to a bolt hole pattern as a child feature.

A functional feature can be considered a shape, reference location or material within a component that serves a specific purpose. An example of a base feature may be a case or a structure, while secondary features may include through holes, flanges, ribs, or passages. Functional features may also include the location of mating surfaces, passages or bolt holes. More and more I am hearing the revised mantra "manufacture for design" rather than the historical "design for manufacture". All well and good until you run up against the laundry list of process constraints and limitations present in all processes including AM.

A manufacturing feature may have no functional end use, but may be added to aid the manufacturing process. An example may be the orientation of a flat surface, a clamping feature, a pad to reduce distortion during welding, a vent hole for a casting and a powder removal hole for internal AM volumes, or simply by adding extra layers of material to a surface to ensure the success of a subsequent machining or finishing operation.

A material defined shape feature is most often associated with the use of a commercially available material stock in your design. In conventional manufacturing, e.g., using sheet metal, the available wall thickness, or surface finish of sheet metal may be considered a material feature. In AM, the minimum wall thickness constraints imposed by the deposition process may also be considered a material dependent feature. Minimum detail size or surface finish may be a function of AM powder material, the model or the process itself. Other features such as an inspection feature, cleaning, powder removal, or post-process finishing features may be considered as well.

9.3.1   Design Freedom Offered by AM

If you were lucky enough to have grown up in a metal fabrication shop you would know about all of the large equipment, how to run it and the tooling it takes to fabricate metal. Take a walk around your shop, or that of your fabricator, and remind yourself what kind of capital equipment, floor space, and store rooms of tooling are required. A foundry will include melt furnaces, casting and mold preparation shops, while a machine shop will feature milling machines, lathes, surface grinders, drill presses, electrode discharge machining (EDM), and boring machines. Sheet metal equipment includes shears, punches, bending brakes, and a host of other tooling as well. AM design may not replace the use or need for much of this equipment, but thinking about what conventional process can be avoided by adopting AM is worth considering. You will see other equipment such as heat treatment furnaces and cutting equipment (such as band saws), polishing and finishing equipment. Which of these will be required to post-process and finish your AM metal components?

If you are lucky enough not to have been born into traditional metal processing and are imaging starting in a shop with just an AM printer and that's all, you should find a metal fabrication shop to tour anyway. Do not forget to visit the tooling storage room, the metal stock room and the all-important scrap metal dumpster. To choose an AM design it is useful to know what is inside the conventional metal processing box and what is outside, what you will need and what you will not. Figure 9.5 shows some of the design considerations.

Here are a few of the design freedoms offered by AM outside conventional processing constraints.

- *Freedom from commercial metal* shape constraints of sheets, pipes, angles, rods, massive billets or blocks.
- *Freedom from conventional process* shape constraints, such as linear bends, straight drilled passageways, square milled corners or sharp edges.
- *Freedom from some conventional post-processing constraints*, such as tool-reach-access if subsequent machining operations are eliminated.
- Ability to *combine multiple parts* into one complex part, reducing or eliminating subsequent assembly and joining such as welding, brazing, soldering or bolting pieces together and reducing flaws and risks of failure associated with other

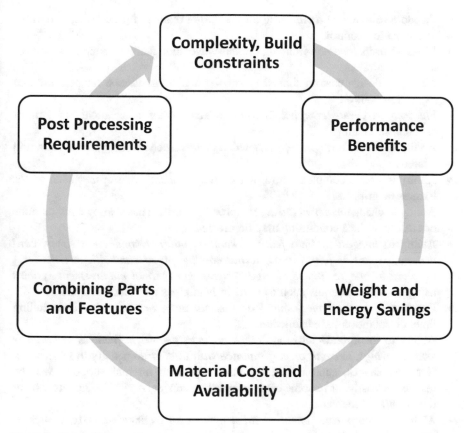

Fig. 9.5 AM design considerations

forms of permanent joints. In some cases plumbing can become integral to the part itself.

- Ability to r*educe waste* material lost by machining or forming scrap (such as punched picture frames). Materials that are expensive, difficult or costly to recycle are good candidates for AM.
- Ability to design *complex internal passageways* such as flow channels for gases, liquids, coolants, or lubricants.
- Ability to *create rigid and lightweight components* by varying wall thickness, eliminating unneeded mass and adding strengthening features only where they are needed.
- Ability to design *tailored* interior or exterior *surfaces*, to optimize thermal or aerodynamic flow, or the adhesion interface with composite materials.
- Freedom to integrate complex linear or rotational features such as *compensators for balance*.
- Ability to integrate *turbulators, funnels, jets, filters, swirlers, strainers or baffles* for mixing or flow or combustion control of liquids or gases.

- Freedom to integrate *dampening and isolation* features for acoustic, harmonic, and vibration control.
- Freedom to include non-imaging optical waveguide features, such as absorbers and concentrators.
- Ability to design custom shaped *reservoirs and accumulators* for pressure and expansion control.
- Freedom to *create complex internal structures* for mechanical strength or anchors for coatings.
- Ability to add *lightweight supports* and *stiffeners*, such as ribs, battens, pads, and gussets.
- Ability to create complex conformal *mating or matching surfaces* such as for fixtures or grippers.
- Ability to change *material during the process*. In some cases alloy compositions or custom surface conditions may be created.
- Ability to *integral thermal features such as internal conduction vanes, condensers, expansion joints, and external cooling fins or bimetallic switches.*
- Freedom to *integral phase change systems and thermal management systems* using multiple materials such as used in heat pipes and thermal siphons.
- Freedom to *integral mechanical or thermal actuators* offering 4D including time, or condition based function.
- Ability to incorporate into the design *complex geometric features to optimize weight savings, strength or performance* with little or no penalty to fabrication. Complex shapes, such as shelled structures with internal supports will be realized as easily, or in some cases easier than convention walls, gussets, ribs or thickly filled sections.
- Ability to *incorporate repeating substructures for engineered designs*, such as linear crush structures.
- Ability to create *porous structures*, such as those used for bone ingrowth in medical implants or filtration.
- Ability to fabricate *structures out of high temperature, hard or difficult to process materials such as tool bits or burner tips*. Complex structures from tungsten or super alloys are two examples of materials that are hard to drill, machine, bend or form.
- *Ability to create individualized and patient matched medical structures* such as cranial or dental implants derived from computed tomography (CT) or a patient's own medical imaging.
- *Ability to design and build on demand components in remote locations* such as in service an ocean vessel or field hospital.

The degrees of design freedom outlined above might assume you are starting from a blank slate. However, if you are designing a part to replace or upgrade an existing part you may ask "does it need to be identical, functionally equivalent, functionally enhanced, modified or customized?" What design, manufacturing or material limitations were present in the original part? Can these be overcome with AM designs

Fig. 9.6 Support structure orienting the part with respect to the build plate[8]

and processing? Is there an existing solid model from which to base a modified AM design or will reverse engineering be needed?

When designing a part one must also consider how it will be oriented during the build and how to design the support structure. Figure 9.6 shows a part with the support structure needed to orient and anchor the part within the build chamber. We provide more detail about this process later in the book.

When replacing a part you may start by considering all the original design features, the purpose of each and how it was made. How did it wear or fail? In essence, decompose the function of each design feature, then imagine reconstruction of design within the newly defined bounds of AM. Design for manufacture takes into account production requirements and processing constraints to assure the former is optimized while the latter is not violated. This philosophy has great payback when designing for large production runs but will also apply to AM production. One must keep in mind the constraints of both AM and conventional methods when designing an AM prototype that may eventually transition into production using conventional means.

---

[8]Credit PSU CIMP-3D, reproduced with permission.

In addition, look into the function of the system containing the component. What is it connected to? Can those objects be combined into one part? Can the function of multiple components be built into one? Small lot sizes for components, requiring brazes or welds, may find economy in consolidated AM designs. Components requiring long changeover time between fabrication steps, or those requiring complex tooling or precision alignments may be candidates for AM. In some cases, support structures upon which AM parts are built and the base plates themselves may be integral to the setup of post-processing operations, such as machining. In those cases removal from the base plate may occur later in the post-processing sequence.

In the book Producing Prosperity (Pisano 2012), p. 114, Gary P. Pisano says

> Sometimes it is much easier to reverse-engineer a product design than figure out someone else's proprietary manufacturing process.

A useful AM design guide is provided at nextlinemfg.com.[9]

9.3.2   AM Metal Design Constraints

Contrary to popular belief, you cannot just build anything your heart desires with AM, although it does open a new world of design possibilities. You can free yourself from many traditional design-for-manufacturing constraints but you may still encounter limitations depending on the material and AM process you choose. But, keep in mind that engineering analysis tools used to simulate heat flow, mechanical movement and fluid flow may be challenged by these new and complex designs. As stated earlier, a large body of in-service performance data does not exist for critical materials or applications so cautious selection of design margins is warranted.

Here are a few design constraints for PBF-L systems

- You cannot just form a free floating part in a powder bed. It needs to be anchored to the build platform to keep it from being swept away at the start of the build by the recoater. Distortion and warping must be constrained to allow each layer of powder to be spread without interference from the underlying fused layer. These anchors are referred to as *support structures*. In addition to anchoring the part to the plate, they also assist in supporting the build of overhanging part features extending beyond the perimeter of the previous layer. Heat sinking may be another function of support structures.

---

[9]Nextline.om design guide, http://offers.nextlinemfg.com/hs-fs/hub/340051/file-1007418815-pdf/Metal_3D_Printing_Design_Guide.pdf?submissionGuid=7460b10f-3fc0-4d29-bbf1-41496e33a133, (accessed March 21, 2015).

- A key to designing support structures is making them rigid enough to withstand the stresses during the build cycle but also fragile enough to facilitate easy removal of the part from the plate and removal of support from the part to allow finishing. Small projections from the support, referred to as *teeth*, are added as an interface between the support structure and the part to further ease removal. Unfortunately, the design of part supports often relies on a trial-and-error approach with experience playing a significant role in a successful design. Parts may need to be designed and built in pieces to allow the removal of support structures prior to joining by conventional means such as welding.
- Overhangs and flat downward facing surfaces require additional support structure. If these are internal to a part volume, the design must provide a hole or clean-out feature to allow access to remove the support material. Part orientation within the build chamber may help reduce the need for supports, but again this requires experience to refine your skills.
- Ports and passages must be provided to remove powder from internal volumes.
- Part orientation within the build volume may affect the quality of the recoat layer, surface quality, and the *stair step* effect on various surfaces at low angle to the build layer plane.
- Minimum wall thicknesses and feature sizes are specified and must be respected for each material and parameter set.
- Maximum build volume dimension constraints may force a part to be divided into sections to be joined during post-processing. Some available software packages assist in the deconstruction of design and creation of parting planes to allow building multiple pieces and subsequently assembling and reconstructing the full design by welding or other joining methods. These methods may include the addition of interlocking or alignment features within the parting joint design.
- High aspect ratio features or rapid transitions in thickness may result in warping or distortions.
- Shrinkage compensation applied to the part design may be needed as derived by trial and error prototypes.
- Post-process finishing requirements may require the alteration of the size and geometry of external or internal features.
- Rapid cooling may affect the metallurgical structure of the deposit and require design or process accommodation.
- To reduce weight and material costs, wall thickness may be thinned and solid objects may be shelled, as long as the required support structures and powder can be removed from internal passages or volumes.
- Spaces between moving parts must be respected and maintained.
- Holes must be sized appropriately for post-process drilling, tapping, or reaming to achieve dimension tolerances.

- Thin sections between thick sections must be appropriately supported or oriented to facilitate the build.
- Higher tessellation resolution may be required to achieve the appropriate surface resolution.
- Corners and sharp edges may need to be avoided to facilitate build resolution, post-process finishing, or specific material/process limitations.

Design constraints for wire feed EB and arc DED systems include

- The large molten pool of wire-fed systems may restrict designs to parts with large sections, avoiding thin sections and small features due to heat buildup and distortion.
- Large melt pools may constrain part building to the flat position preventing the deposition of overhang features and limiting walls to vertical or decreasing cross-section features.
- Passageways with ceilings or overhang features may be impossible to fabricate without process interruption and placement of inserts, ceilings, or runoff tabs.
- The designer will have to account for the potential for large distortions and residual stress related defects such as hot tearing or cracking.
- Cooling of very large or massive parts will proceed slowly in a vacuum chamber due to reduced thermal conductivity within the vacuum. This may lead to the undesirable grain growth within the deposit.
- Long cooling times may lead to the undesirable diffusion of interstitial contaminants, such as hydrogen gas into the bulk material.
- Mirror image deposition with respect to the base plate may be required to offset distortion during cooling. The addition of restraint features may be needed to control distortion during the build, to be removed during post-processing perhaps before or even after heat treatment.
- Ribs and other support structures may need to be added to support the build of large flat sections or thin sections susceptible to warping.

Design constraints for DED-L systems include

- 3–5 axis deposition may require tilt angles or head access not achievable by the deposition head.
- Clearance of the powder and laser head may dictate positional orientation of the build as well as part geometry considerations.
- Decreased resolution and edge retention may further constrain corner and edge geometry.
- Knowledge of the center of gravity or mass may be required if the part is being articulated.
- Fixture attachment points, clamping, or positioning locations may be needed for the base plate or base feature.

- Integration of the base feature into the final design may require the designer to create or reverse engineer an existing 3D base feature model.
- Certain geometric features may not be buildable when compared to PBF-L designs.
- In hybrid systems, a selection of AM deposition heads may require a range of parameter sets developed for each feature being fabricated, thus complicating the CAM design process and requiring a high level of human expertise and manual programming.

At this point you have considered all the applicable constraints and tradeoffs and have completed your part design. You have created a solid model, and in the case of powder bed systems, have checked the STL file and fixed any errors. Part orientation and support structures have been designed and positioned within the model of build chamber.

Directed energy deposition processes follow a similar path from model to CNC deposition path with all machine function codes embedded into the control file. When using milling or turning CAM software, simulation of the tool path and material addition sequence provides an extra level of confidence that you have error-free motion control and will end up with the desired shape. In the next chapter, we will describe the actual build of a prototype or production part.

9.4   Additional Design Requirements

9.4.1   Support Structure Design

Consideration of support structure design (Fig. 9.7) is a critical step for the design and fabrication of an AM components. Figure 9.8 shows both parts and support structures built within a single build volume. Supports are typically required for powder bed type processes using lasers and to a lesser degree with PBF-EB. Base plates, base shapes or structures may also be used for DED and hybrid systems. An electron beam process, such as Arcam's EBM process may partially consolidate powder adjacent to the part being built eliminating the need for specifically designed support structures.

Unfortunately, the design of support structures may be a trial-and-error process relying on experience and skill to optimize the process. Building support structure may account for a significant portion of the build time and materials and if they fail, what otherwise may have been a perfectly good part will be scrapped. Support

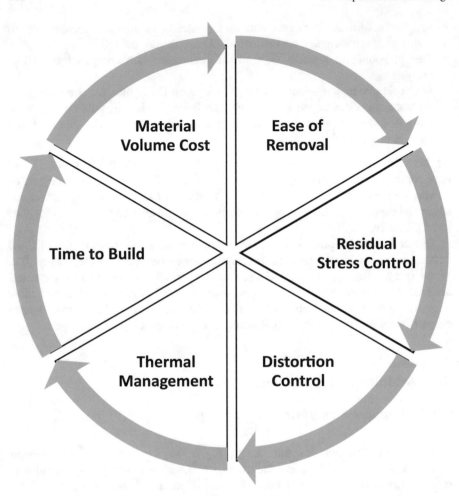

Fig. 9.7 Support structure design considerations

structures and part orientation may need to be considered carefully to avoid shrinkage stress generated during the build from peeling or delaminating the part from the base plate.

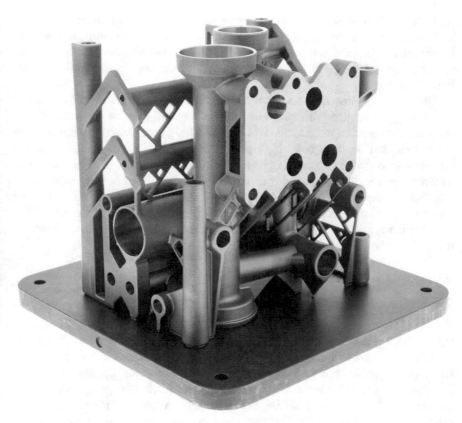

Fig. 9.8 Part and support structure shown within a single build volume[10]

In an article in Modern Machine Shop,[11] the same CAD model was sent to three different AM service providers to use their PBF-L machines to fabricate a part designed with features relevant to turbine components. Unknown to one another, all three independently chose to orient the part in different ways and all three produced different results.

9.4.2 Design of Fixtures, Jigs, and Tooling

Post-processing of AM parts may often require the design and fabrication of fixtures, jigs, and tooling to enable operations such as drilling, tapping, reaming,

---

[10]Courtesy of Renishaw, reproduced with permission.

[11]Modern Machine Shop, http://www.mmsonline.com/articles/additives-idiosyncrasies, (accessed March 21, 2015).

machining, EDM, welding, engraving or inspection. Fixtures for precise handling, such as grippers for robots, may also be required. These manufacturing aids may be generalized to be used across a family of parts but may also be tailored for individual components. The same optimization criteria may be applied to the fabrication of these parts.

Single use fixtures may be fabricated integral to the parts to allow positive positioning during post-processing and may be removed later, or they may employ adjustable features for multicomponent or multi-design use. Metal is often used for these parts, but in some cases high performance plastic may be used. Surfaces conforming to the actual part may be used to fabricate custom grippers, handling, transport, or storage aids. Complex ergonomic design features used to assist human interfaces to the process may offer handling, gripping surfaces, reach or balance features. The cost of an AM-produced fixture must be compared to the cost of less complex clamps or jigs. Just because you can make it using AM does not mean that it is the best way.

9.4.3   Test Specimen Design

Partial fidelity test specimens and testing are often required as part of process development. Test specimens can be limited in size and shape to economize time and material usage, but they must accurately represent the portion of the part function being tested. One example would be the fabrication of cylindrical bars machined into standardized tensile test specimens. Another example is the fabrication of a partial portion of the part geometry such as a small feature or passageway to determine the ability of the process to deposit a difficult or small feature. The potential exists for AM to deposit an entire array of standard mechanical or metallurgical test specimens at once, thus reducing the time for individual sample preparation. All samples may be deposited in a single build sequence, at the same parameters or over a range of parameters, thus exploring or characterizing the process parameter space.

9.4.4   Prototype Design

Prototyping has historically been a slow and costly process. Prototyping is used to evaluate and optimize design and develop fabrication procedures. In addition to confirming the design, prototyping may be used to develop nondestructive or destructive testing procedures. Failure to meet the requirements at this stage may result in the need to go all the way back to the original model design and make changes based upon what was learned. In other cases, a simple change to the process parameters may be all that is required. The design and fabrication of custom tooling and fixtures often accompanies redesign iterations, compounding costs, and delaying first article production. The first prototypes may be made of plastic to determine form and fit while later iterations are produced in metal to test functionality. Rapid prototyping, offered by AM, refines the design and speed the redesign process in the building and testing of parts, ultimately producing a better final component. The concept of de-risking final design is realized by employing a faster and more robust prototyping process.

9.4.5   Hybrid Design

Hybrid designs may be realized in different forms. One example is the incorporation of additive design features along with subtractive design features, produced on a hybrid or retrofitted AM/SM processing system. Another example results when an existing commercially shaped base part is modified with the addition of additive features to form a hybrid produced final component.

Commercially produced hybrid machine tools that integrate both CNC machining and AM deposition capability are being introduced. To date, this hybrid integration has been mostly limited to CNC/DED systems. As with CNC machining, design begins with commercial metal stock, creating subtractive and additive features in sequence to realize the final shape. In-process dimensional inspection, or other operations such as localized heat treatment, glazing or finishing may also be integrated into a single platform. Combining and adding AM part features with commercial shaped stock can speed the fabrication of a final shape by leveraging the cost and processing benefits of beginning with simple commercial shapes.

As introduced above, another hybrid application is the use of modular design to accommodate limits in build chamber size by splitting up or sectioning a component design into small subcomponents to accommodate the confines of a small build chamber. Novel joint design features may be added to accommodate subsequent assembly and joining, such as lap joints for conventional brazing or welding or interlocking joints. In one example, a bicycle frame is constructed with AM-produced lug joints, custom designed to fit the size and dimension of a bike rider. Straight lengths of commercial shape tubing are then cut and joined to the lugs, resulting in a custom-sized bike frame.

9.5   Cost Analysis

Cost estimation, business consideration, and the economics of AM require consideration all along the design and process development cycle (Atzeni and Salmi 2012) (Frazier 2014) (Sames et al. 2015). Figure 9.9 identifies some of the cost considerations when designing for AM. Figure 9.10 shows a representation of a typical cross-over break even analysis based on the Deloitte break even analysis

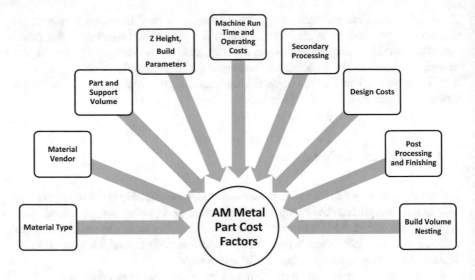

Fig. 9.9 Cost factors to consider when designing for AM

approach[12] where the cost to produce a part by conventional methods is compared to the cost of producing an AM fabricated part. Conventional part production requiring significant upfront costs associated with tooling and process development along a long production stream must rely on production quantities and sales to recover the upfront costs and be profitable. AM part production can reduce upfront tooling costs and make sense for small lots of expensive parts. It is important to note a learning curve for producing parts by AM does currently exist and may require a number of iterations before a successful first part is produced therefore increasing the cost of the first part produced. Another important consideration regarding the adoption of AM methods is the current uncertainty with total costs of AM parts, materials, design, and fabrication. It is important to note the prices range widely with respect to powder quality, quantity, and purity. A rule of thumb is to expect wire products to cost twice as much as commercial shapes and powder products to cost twice as much as wire products. However the price for AM powder feedstock will drop with greater AM adoption. With experience, greater adoption and competition within the market, the region of profitability for AM production will be more clearly defined. If you are planning to buy an AM machine and establish an in-house capability, the table below lists some of the cost considerations.

---

[12]3D opportunity: Additive Manufacturing paths to performance, innovation, and growth, Deloitte Review issue 14, Mark Cotteleer, Jim Joyce, January 17, 2014, https://dupress.deloitte.com/dup-us-en/deloitte-review/issue-14/dr14-3d-opportunity.html, (accessed January 28, 2017).

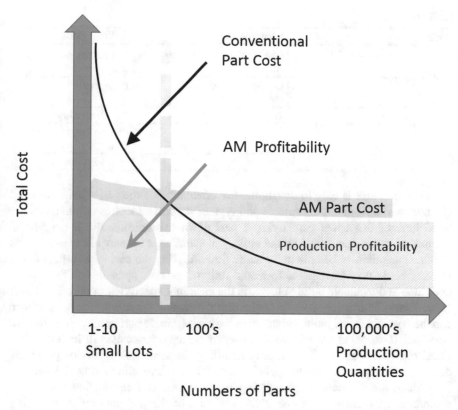

Fig. 9.10 A typical break even analysis based on the Deloitte break even analysis approach

In buying an AM metal printer you may well spend a million dollars getting a shop set up, purchasing the machine and getting it up and running. The initial investment in materials can add up if you plan to use the more expensive alloys: you may spend up to tens of thousands of dollars on metal powder stock alone, depending on your selection. If you plan to build your own parts or parts for others as a service, there are additional costs to consider, such as machine purchase cost recovery, machine operator costs, material costs, consumables such as inert gas, utilities, depreciation, maintenance, repair and the useful life of the machine. The cost of quality, documentation generation and retention requirements all come into play.

Fixed Costs	Recurring Costs	AM Part Costs
Capital costs	Hardware Maintenance	Part design cost
Initial facility and installation costs	Software Maintenance	Tech and touch labor cost
Amortization		

(continued)

(continued)

Fixed Costs	Recurring Costs	AM Part Costs
	Consumables, build plates, etc.	AM metals, inert gas, etc.
	Engineering and facility support	Energy cost
	Training	Post-Processing
		Prototyping, trial and error
		Failed Build

While part size is easy to determine, the orientation of the part within the build chamber will affect both the volume and type of support structures required and the build time of the actual part. Lying a part down to minimize the Z height and number of layers may speed up the build, but standing it up may allow more parts of other customers to be built at the same time. Angling the part with respect to the recoater blade may be required to assure uniform powder spreading at the cost of both time and available build volume. Is the customer specifying build orientation? Other factors to consider include single unit fabrication verse batch costs, part size and the cost of a failed build, either when building a single part or a batch. Is rework possible? If so, what are the costs? Material changeover costs will include additional chamber cleaning and material handling. In some cases, service providers dedicate specific machines to specific material therefore eliminating the need to fully clean the chamber when building the next part and eliminating the risk to contaminate the material of one build with that of the previous build. Finishing costs will be a function of material support removal and a wide range of options selected by the customer. Additive Manufacturing Technologies, Rapid Prototyping to Direct Digital Manufacturing (Gibson 2009, p. 374), describes a cost model that takes many of these parameters into consideration.

If you plan to have parts built by a service provider, does the provider have multiple machines dedicated to specific materials? Do they have machines with build volumes that suit your component size? Will you be paying a premium to have a small part build in a large build volume? Will they choose the best orientation to achieve the best accuracy or will they lie it down to reduce the build height and build time? Will your part be built at the same time with other customer parts with at risk designs? Will they build support structure on top of your part to stack fill the build volume? Will they be waiting for other customer orders to fill a build volume? If the process is interrupted and restarted will you be informed and provided the restart data? Assuming you are on a rapid prototyping schedule, what are the costs of those delays to you? How experienced is the provider with AM of those materials? If you order the same part a month later will it be the same? These are all good questions and certainly not a complete set.

9.6 Key Take Away Points

- Artistic and engineering applications approach design from opposite directions beginning with either the desired form or function. Additive manufacturing methods and processes, correctly applied, can provide advantages to either approach.
- Complex designs or shapes that cannot be fabricated by conventional methods are made possible by AM.
- Effective AM designs must incorporate an in-depth knowledge of the materials, the specific AM metal process, support structure and other processing or post-processing operations to settle on an effective design.

Chapter 10
Process Development

Abstract The power of AM metal processing is realized from both the design freedom and the removal of many of the constraints imposed by commercial shapes (such as flat sheets, round pipes, etc.) and some of the constraints imposed by conventional processing (such as straight drilled passages, linear bends, etc.). This freedom comes at the cost of process complexity as the many process parameters available to the AM engineer are often poorly understood and hard to control. While AM machine vendors will offer parameter sets for a specific set of materials and often offer design and process consultation, AM users are often faced with designing and optimizing a prototype process along with designing and optimizing a prototype part. This chapter steps the reader through AM process development and the process of parameter selection to optimize part density, surface finish, accuracy, and repeatability while reducing build time, distortion, residual stress, other defects and conditions that may affect part quality or performance.

10.1 Parameter Selection

Having chosen a design, material and process, process development, and parameter selection comes next (Fig. 10.1). Process parameters are those machine settings that produce the desired deposit, meeting part requirements, overall quality, and repeatability. The leading AM machine builders have developed standard parameter sets to work with their machines and proprietary powders. They have in some cases spent years developing these parameter sets for each of the materials they offer and in many cases hold this information as proprietary. These parameter sets produce quality, near-full-density deposits when built using the AM machine vendor's recommended conditions. After ascending the learning curve for component and support design and placement, these parameters can offer a high degree of confidence in the production process, requiring little user intervention. For companies wanting to do their own process development, perhaps using their own specialty powders or proprietary designs, users can find themselves locked out of the process

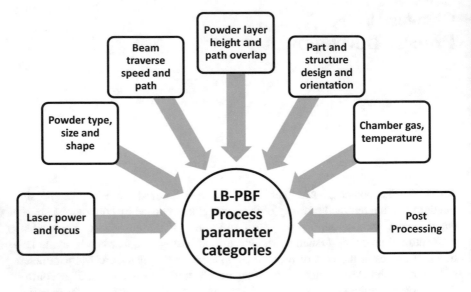

Fig. 10.1 Laser PBF parameter categories

development or process improvement cycle. In these cases, your in-house developed IP and process knowhow is dependent upon the proprietary IP of others.

Standard parameter sets supplied by vendors may be protected by encryption techniques that prevent the user from seeing the actual parameters or modifying them to suit in-house development needs. You can expect to pay a premium for a hardware key that enables the use of standard parameters for each material of interest. If you expect to be running multiple materials, you can expect to spend large sums to obtain each license set for each material type. Other detrimental aspects of proprietary parameters or system architecture are difficulties encountered during process qualification or certification of standardized procedures. If you are not provided the primary process parameters used to build the part, how can you assure they are the same or equivalent across software upgrades? The same question may be asked when expanding your production to a large base of upgraded or next generation AM machine types.

Some machine builders are offering open architecture and software platforms that remove the constraints of proprietary materials and parameter sets. Open architecture can free the user to do their own in-house process customization, integration, and development. However, the complexity of AM processes and the time and effort required to develop your own process parameters can be costly and time consuming, therefore careful consideration of the pros and cons is required when setting out on your own.

AM machine types feature a wide variation in hardware, software, and function. In this chapter, we will provide a look into what type of parameters can be adjusted, which can be monitored and controlled and which are held constant. This chapter then describes the process of parameter selection and procedure specification without focusing on any one type of machine.

Trial and error development of process parameters may be required to optimize the deposited material to minimize defects and meet the desired end result. With experience and good records, you may ascend the learning curve for your materials and part types, reducing the development cycle to simply a verification of an existing parameter set. Earlier in the book, we described the primary process parameters for laser, EB, arc, and plasma arc heat sources. We also discussed the specification for powders and filler wires typically held as constant. In review, heat source parameters included heat source type, power and focus condition. Powder parameters include chemistry, particle size distribution, and shape. As an example, changing to a different powder vendor, with a different powder specification, may require other changes to primary process parameters. The difficulties of AM process development are well known in industry and are one of the barriers to entry for wider adoption of the technology. Efforts are underway to create databases and computer software to assist in process parameter selection. As an example the US government program America Makes[1] has granted General Electric (GE) and Lawrence Livermore National Laboratory (LLNL) funding[2] to develop an algorithm to help choose and adjust primary SLM parameters based upon materials.

Process parameters for the relative motion of the heat source to the work piece include scan and traverse speed. Sophisticated scan paths or scan strategy developed by vendors include alternating scanning locations within a layer. Scanning with a square or band shaped region (Fig. 10.2a, b), rotating or randomizing scan orientations from layer to layer will reduce the buildup of thermally induced stresses, distortion and to promote full fusion of the deposit. Leveling, smoothing and improved surface condition is another reason for altering the scan path orientation between layers or along the layer edge contours or perimeter outer surfaces.

Scan conditions may change if the deposition path is a contour or fill type region. Parameters related to material feed can include powder layer thickness, powder feed rate for DED-L, and wire feed speed for DED-EB or arc-based systems. Direct metal deposition using multiple wire or powder feeders may specify a mix ratio, as a function of location within the part, when performing graded metal deposition. Inert gas parameters will include type, purity, flow rate, purge time, or in the case of DED-L, powder feed gas, flow, and pressure. Preheat or post-build cooling times are among the thermal conditions considered as adjustable parameters. Layer-to-layer changes to parameters for PBF can include the number of contour passes or the rotation angle of raster paths. This is not a complete list of all the degrees of

---

[1]The US government America Makes program is funding a wide range of AM technology development bringing together universities, corporate leaders and small businesses, https://americamakes.us/, (accessed March 22, 2015).

[2]LLNL news, America Makes taps Lawrence Livermore, GE to develop open source algorithms for 3D printing, March 13, 2015, https://www.llnl.gov/news/america-makes-taps-lawrence-livermore-ge-develop-open-source-algorithms-3d-printing/, (accessed May 14, 2015).

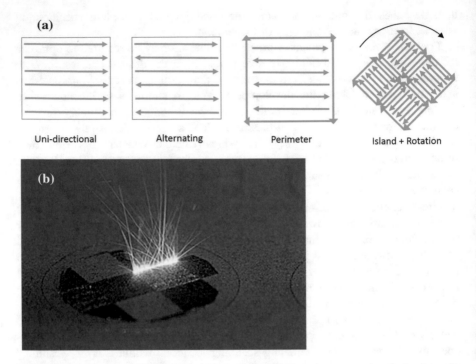

(a)

Uni-directional   Alternating   Perimeter   Island + Rotation

(b)

Fig. 10.2 a Scan paths and scan strategies for PBF, **b** A laser beam scanning the surface of a powder bed based upon a predefined path[3]

freedom these parameters offer the process, but as you can see, there are many and they all make a difference.

As described above, the number of parameters available and the range over which each may be adjusted or changed leads to a highly complex set of choices. After you understand the number and range of parameters available to your process, the question becomes which combination of all of the above will optimize your bulk material deposit, build efficiency, meet part requirements, and ultimately optimize part performance. This is not an easy task given that many of these parameters interact in a complex and nonlinear manner.

If you are developing your own parameter set, where do you start? Hopefully if you have a commercial system that allows adjustment, you start with their base set, if they let you see it. If you have an open-source machine, you may be able to interact with a working group of collaborators and use their baseline settings. Otherwise, trial and error may be your only option until open-source databases are constructed. For sintered metal optimization you may focus on maximizing deposit

---

[3]Courtesy of Beamie Young\NIST.

density or build rate. For DED systems you may focus on distortion control and part accuracy. Trial and error, when used in selecting the huge number of process parameters, can be an inefficient and resource consuming task. Until there is an AM Handbook analogous to the Machinery Handbook where you just pop it off the shelf to look up speeds, feeds, cutting tool selection and rake angles for different materials, you are at the mercy of a new technology.

Power and travel speed are two such conditions and are often referred to as *primary process input variables*. They may be set ahead of time and held constant throughout the process, or as under closed-loop control they may be changed and controlled during processing either by the operator or by an automated control system. Understanding the role and relationship of these primary input variables to control sintering or fusion is critical to understanding how to achieve the desired density and properties of deposited material. These fundamentals of melting and solidification apply to all heat sources and molten pools but the nature of those processes at very high speeds and very small molten pools can make them difficult if not impossible to observe. In welding related processes changes in travel speed and welding current will result in changes to the bead shape, width and penetration depth. The same relationship is seen in AM metal processing but at a much smaller scale. Describing this relationship in weld process terms is much easier for most students to grasp.

Other parameters referred to as *secondary process input variables* are often held constant during the process development cycle. Inert powder delivery gas or build chamber gas selection such as argon or nitrogen are examples of secondary input variables. As an example, while the thermal conductivity of nitrogen is 40% greater than argon and can change chemistry of cooling rate effects, the gas composition is held constant effectively eliminating it as a variable.

Primary process output variables are what the AM operator may see during processing or after the build process by inspecting the deposit. Melt bead profile or depth of melt penetration are two examples. Common AM output variables can include surface condition or deposit density.

Secondary output variables, such as those caused by disturbances may include flaw generation or defect formation. Other output variables may include stresses created by thermal expansion, thermal softening, distortion, and shrinkage. The operator may not see problems such as bending, buckling, curling, delamination of tearing until the build fails (Fig. 10.3).[4] These effects are hard to predict, observe, and control but may affect the part quality. It all depends upon the specification of the part. With an understanding of the moving heat source, part fixtures, support design, process selection, and with rigorous development, these problems may be controlled or reduced to acceptable levels.

---

[4]NIST Technical Note 1823, Additive Manufacturing Technical Workshop Summary Report, Christopher Brown, Joshua Lubell, Robert Lipman, http://www.nist.gov/manuscript-publication-search.cfm?pub_id=914642, (accessed May 14, 2016).

Fig. 10.3 A standard test article torn from the build plate[5]

10.2 Parameter Optimization

By now you either have a parameter set based on vendor supplied standard conditions, are starting with someone else's parameters, or are starting with a blank layer of powder. If the latter is the case, you must develop your own parameter set. You need to start with enough beam power at a travel speed to begin melting and fusing the powder to the build plate or to a support layer. Once you have achieved this you will want to optimize the deposit conditions. Here is a listing of some of the conditions and other related process considerations.

Bulk Deposit Density

In general the more power you apply to a localized heat source, for a fixed travel speed, the larger the molten pool or sintered region. A faster travel speed and a fixed heat source power will result in a longer and narrower melt region until melting is no longer achievable. There are, however, limitations, as both the power and speed can exceed the stability of the process, resulting in loss of a smooth continuous deposit or complete loss of melting. A near full density deposit is generally desired to achieve the mechanical properties desired in engineering alloys. A continuous molten pool provides the best conditions for a fully dense melt evolved microstructure. The tradeoff for increased melt pool size is that of excess heat input may result in the

[5]*Source* "Materials Standards for Additive Manufacturing," John Slotwinski. Appendix 4 in: Additive Manufacturing Technical Workshop Summary Report. Christopher Brown, Joshua Lubell, and Robert Lipman. NIST Technical Note 1823, National Institute of Standards and Technology, Gaithersburg, MD. November 2013. doi:http://dx.doi.org/10.6028/NIST.TN.1823.

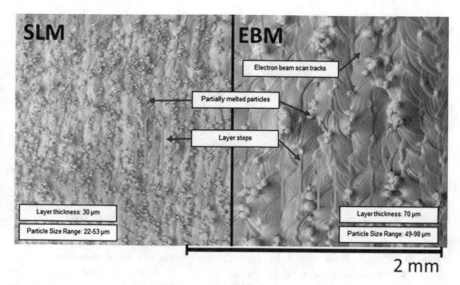

Fig. 10.4 As-deposited surfaces of SLM and EBM[6]

formation of solidifications stress, distortion and the loss of deposition accuracy. In some cases, a deposition schedule may require multiple passes and scan orientation paths (crisscross or rotation) to assure full fusion of underlying layers. If your power density is too low you will not achieve the desired degree of melting. If it is too high, you may transition into a keyhole mode of melting resulting in an unstable process, voids, cold shuts, spatter, and excessive vaporization.

Reduced Build Time

The time required to build an object is constrained by a number of factors. The resolution of the build sequence is a major factor. How thin are your recoat layers? How much overlap is required between each scan pass? The distance between tracks or raster patterns is tied directly to beam power and build speed. If there is too much distance between scan paths you risk not fully fusing one track to the next. Too little distance between tracks and you risk excessive heat buildup and excessive build time. Other factors include the size, weight, and inertia of the mechanical systems used to spread the powder or articulate the material deposition hardware. Some DED systems articulate or move the part, others move the torch or laser head, powder, or wire feed systems. Others reduce these constraints by articulating laser motion by simply moving the optics to scan the beam within the build envelope of the system. Ongoing R&D and process development by OEMs

---

[6]*Source* Surface texture measurement for additive manufacturing, Andrew Triantaphyllou et al., Published 5 May 2015, Surface Topography: Metrology and Properties, Volume 3, Number 2, http://iopscience.iop.org/article/10.1088/2051-672X/3/2/024002. © IOP Publishing. Reproduced with permission. All rights reserved.

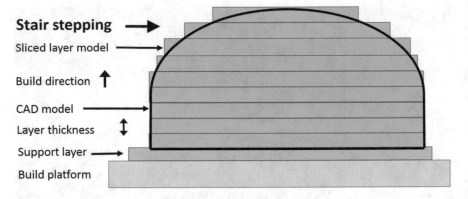

Fig. 10.5 Stair stepping as a function of surface angle in relation to the build plate

looks to increase build speed by incorporating multiple laser scanners or heat sources and to reduce the time to spread a powder layer or further automate the pre-build and post- build sequence steps.

Surface Finish

Objects made from fused powder have a surface finish of partially fused powder and display tracks of the layered melting, as shown for both SLM and EBM built parts (Triantaphyllou et al. 2015), in Fig. 10.4. To one degree or another, surface variations typical to AM deposition of one track fused to the next and one layer built on top of another will be evident in an irregular surface condition commonly referred to as stair stepping (Fig. 10.5). Surface roughness for SLM/PBF-L may be in the range of 300–600 μin. and 800–1000 μin. for EBM/PBF-EB.

Surface characteristics resulting from fused wire deposition will display very large overlapping beads of materials in stair stepping patterns characteristic of the type of material, wire size, and deposition conditions. A large melt pool with corresponding high deposition rates will result in coarse deposition features and surfaces. The fluidity of metals can vary widely and will change the shape of the molten pool and the final deposit. This will also affect the surface condition. Realizing all of this, manufacturers are offering smaller powder diameter sizes enabling the use of smaller powder recoat layers. However, these optimizations always come with a tradeoff. Parameter conditions for contour or perimeter passes may be altered to allow finer detail, smoother surfaces and decreased stair stepping, at the cost of build time. Hybrid AM systems relying on DED now offer two laser head deposition types, one for fine detail and one for faster deposition. Smaller powder size may complicate powder handling or present additional hazards as described earlier.

All conventional metal processes and materials have their own surface characteristics that are accommodated in various ways depending on the application. Cast surfaces may require machining, polishing, or sealing with a coating to achieve the desired performance. Hot-rolled steel or welded structures may need finishing to

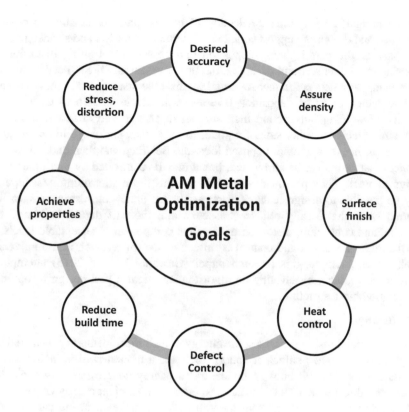

Fig. 10.6 AM process optimization goals

remove mill scale and rust or to modify surface conditions. AM surface conditions may be partially optimized by parameter selection or they may see additional optimization during post processing. The designer and process engineer must understand the nature of achievable surface conditions for a given material and choose the best method to optimize the desired end result. Part models dimensions may need to be offset to provide additional material to be removed by machining or polishing, often called *machining or finishing allowance*. Additional optimization goals for AM parts are shown in Fig. 10.6.

Accuracy and Distortion

Build accuracy is affected by a number of conditions, such as part design, powder type, size and morphology, path resolution and processing parameters. As you recall from earlier in the book, thermal expansion and contraction is a property unique to specific metals and alloys as are other properties that affect shrinkage and distortion. Shrinkage compensation may be built into the computer model or effected by orientation of the part within the build volume. Shrinkage or distortion may be mitigated by optimized selection of process parameters, such as preheat, post-heat, and design of support structures. Thermal expansion and shrinkage come into play as the part is being deposited and cooled. Controlling distortion by choosing the proper restraint during fabrication is essential. Direct metal deposition systems may use deposition pathsoriented to offset the distortion of a previous pass. This staggered offset of deposition layers or scan tracks is akin to the common practice of welding one section then another in an opposed orientation, such as alternating welds on either side of a plate, to help offset shrinkage in one location with the shrinkage in another opposed location. Bending, warping, and curling is a problem that can never be eliminated, but it may be controlled by part orientation, design geometry, and parameter selection, such as those controlling heat input. It may be possible to design features, such as shelled honeycomb structures, or thin internal walls that locally relieve stresses within the part during building, thus reducing and controlling distortion of the final component to acceptable levels.

Distortion prediction and control is an active area of research at the university level. As an example, this research paper (Giovanni Moroni 2015) attempts to address the problem by creating an algorithm to orient a part design to optimize accuracy within a structure.

Heat Buildup

As with all metal fusion processes, the energy to melt most metals is substantial and the process to localize sufficient energy to create a molten pool is inefficient. As discussed earlier in the book, a concentrated energy source (arc, laser, electron beam, etc.) directed at a metal surface experiences inefficient energy coupling and melting. Some of the energy from the source never makes it to the part, some is reflected away and only a percentage is absorbed into the material. Of the energy absorbed, some is reradiated back out of the material and lost, some is conducted into the part and if sufficiently localized, will raise the temperature of the material to create melting. Heat energy will continue to be conducted away from the melt region into the part, raising its temperature. Depending on the size of the part, heat buildup may sufficiently raise the temperature, leading to undesirable effects such as surface oxidation and other chemical reactions. These reactions may adversely affect the surface chemistry or microstructure, such as with impurity segregation, excessive grain growth, overaging or unwanted softening. As stated above, localized heat buildup may also result in unwanted distortion or residual stress, which may lead to cracking or other part performance issues. In some cases of PBF, the degree of melting and the amount of energy absorbed as a function of time results in negligible heat buildup and may not be an issue. In other cases, preheat or maintaining high powder temperature during the build cycle can reduce these effects.

The goal here is to achieve melting and fused deposition when adding material, be it powder or wire, while limiting overall heat buildup or by reducing the difference of the localized heat conditions of the molten pool region and that of the overall component. Starting with the heat source, a highly focused heat source such as a laser or electron beam is more effective in creating the desired degree of melting while limiting excessive additional heating of the part. A less focused heat source, such as an unconstrained GMA welding arc, heats a wider area adjacent to the melt pool therefore depositing unwanted heat into the part.

The heat source is often used to melt the filler material into the base metal. Another way to melt the filler is by using the energy in the molten pool to melt the filler. This method is less efficient from the heat buildup stand point as the filler tends to quench the pool, requiring more energy to be coupled into the pool. Commercial welding systems can utilize a wire preheater to increase deposition rates and reduce heat input into the part. Efforts are underway to provide the same filler preheat to DED-L AM.[7] Heating of the powder bed for PBF processes can accomplish the same effect by reducing the intensity of the high-energy heat source and reducing the thermal gradient near to the melt region with the ultimate goal of higher deposition rates, less distortion, and residual stress.

Other schemes to reduce heat input, while maximizing deposition rates and reducing distortion, may include hot wire PA-DED. Variable polarity plasma arc may be able to bias the polarity to direct more energy in the reverse polarity portion of the waveform to assist in filler metal melting, while minimizing the straight polarity portion of the waveform energy to create only enough surface melting to ensure fusion of the deposited filler.

Increasing travel speeds and power levels can also reduce total heat input per inch of deposit, maintaining adequate fusion to the substrate and adjacent deposit. At some point the beam energy density will transition from conduction mode melting to keyhole mode melting, creating a vapor cavity which may be detrimental to the deposit quality, leading to porosity, entrapped voids, increased spatter, and vaporization. This can have a detrimental effect on the part, recycled powder supply and increase undesirable condensed metal vapors within the chamber, on the laser window, or within the electron gun.

Tool path planning and optimized dwell times may be used to avoid heat buildup in fine features or those thermally isolated from the bulk part offering heat sinking. Complex design features such as heat dams, cooling fins, or other features may be specifically designed into the part or support structure to provide functional enhancement to control heat or stress buildup during the build.

Mechanical Properties

Optimizing mechanical properties of melted or fused metal is often achieved by controlling density of the deposit, minimizing defects and the controlling the cooling rate within the fusion and heat affected regions. Understanding and controlling defect generation, microstructural evolution during solidification and solid state transformations upon cooling are needed to optimize and control the properties of the deposit (Frazier 2014; Herzog et al. 2016; Sames et al. 2015). As stated

---

[7]America Makes, funded research project, High-Throughput Functional Material Deposition Using a Laser Hot Wire Process–Case Western Reserve University (4032), in partnership with Aquilex Corporate Technology Center (AZZ, Inc.); Lincoln Electric Company; RP+M, Inc.; and RTI International Metals, https://americamakes.us/home-2/item/501-high-throughput-functional-material-deposition-using-a-laser-hot-wire-process-case-western-reserve-university, (accessed March 22, 2015).

earlier in the section on sintered metal properties, understanding, and controlling the degree of sintering as a function of location and parameters is key to controlling sintered metal properties. Determining the mechanical properties is a long and evolved process utilizing test sample fabrication and extensive testing and characterization to accepted standard testing methods. One of the big problems associated with AM results from the layer wise deposition and the fact that the mechanical properties of the deposit will vary in relationship to the test sample orientation within the layered structure. Heat treatments and HIP processing has been shown to reduce or effectively eliminate these anisotropic properties, but much more work needs to be done for all materials and all machine configurations. Without extensive characterization of a specific process, material and parameter set the best option is to refer to the mechanical properties supplied in the vendor-supplied technical data sheets. The good news is that vendor-supplied data and parameter sets can achieve or even exceed those of conventional material forms. The bad news is the lower bounds of material properties as a function of poorly developed process parameters or defects is poorly defined.

Chemistry and Metallurgy

Surface chemistry optimization may be as simple as controlling build chamber purge times prior to a build, or using increased dwell times upon cooling, in an inert atmosphere, to avoid discoloration. Selection of the inert gas quality or following pressure cylinder change-over procedures to avoid gas line contamination may need to be specified and controlled within the build sequence. Inert gas purification and recirculation systems can have a significant effect on the chemistry for certain materials. Other procedures, such as the sieving, handling, and bake out of reused powder, may also affect the chemistry and metallurgy of the final deposit. Controlling bulk chemistry and metallurgical effects involves prediction, sensing, and control of grain growth, phase transitions, and other metallurgical processes. This is an active area of research at many universities and national laboratories to incorporate modeling, simulation and data-driven materials development and has recently been referred to as identifying the Material Genome.[8] Directional solidification, single crystal growth and the creation of complex intermetallic phases are beyond the scope of this book but AM applications, particularly those involved with aerospace materials, exist in all of these technical areas and may be researched by the reader. Resources such as Google Scholar are a good place to start the search.

Defect Control

Parameter optimization is needed to reduce or eliminate discontinuities or defects, at all locations within the part. Sensing and control of AM defects is the subject of ongoing industrial R&D. Knowing the type of defects, how they are formed and how to sense and control their formation is key for achieving overall process quality

---

[8]Materials Genome Initiative, http://www.nist.gov/mgi/, (accessed March 22, 2015).

and process certification for certain critical applications. Detection and characterization of these defects is described in later in the book under the sections for destructive and non-destructive testing.

Preparation of Test Coupons

As mentioned above, parameter optimization is evaluated and iterated through the fabrication and testing of samples and early prototypes. Simple test samples, fabricated as small coupons, may be used to evaluate parameter sets against performance requirements. Some of the commercially available software packages allow the assignment and marking of identification codes, or data matrix ID's for each test sample fabricated to specific parameter conditions. Post processing conditions may be evaluated as well, verifying machining allowances, polishing or other operations, such as heat treatment. This is done through nondestructive and destructive methods, as described in more detail below. Visual inspection and metallographic evaluation is often employed to direct parameter selection. Critical applications may require the fabrication of tensile or mechanical testing specimens if functional prototypes are planned. A down-select is then performed to identify the most promising build schedule to be used during the full prototype stage of development.

Evaluating Prototype Performance

Test samples may be used to assure successful selection of candidate parameters, enabling the building of partial fidelity or functional test quality prototypes. Full metal prototypes may also be used for development of nondestructive test procedures, such as dimensional measurement, surface characterization and other nondestructive or destructive test procedures to characterize the bulk deposit morphology in multiple regions of interest within the component.

Upon successful inspection, parts may be subjected to further testing to confirm functional fitness. Structural test parts or finished components may be tested by simulation under actual service loading conditions, while pressure or vacuum leak testing may be appropriate for storage vessels. Test results may indicate the need for redesign, rework or modification of process parameters, or in the event of a failure, back to the drawing board in the redesign phase.

10.3   Specifying Pre-build and Monitoring Procedures

At this point you have designed your part model, support structures, post-processing fixtures, test coupons and any early prototypes. You have chosen your material and one or more sets of candidate build parameters. If you are using a powder bed system you have converted your files to STL and fixed any bugs or issues. If you are using CNC control you have created a CAMor CNC control file and performed dry run simulations of the deposition path to assure smooth motion and collision avoidance. Before beginning the build cycle, there are pre-build, build monitoring, and post-build procedures to be documented.

The National Institute of Standards Technology (NIST) technical note NIST1801, "Lessons Learned in Establishing the NIST Metal Additive Manufacturing Laboratory" (Moylan et al. 2013) does a good job of identifying pre and post-build procedures for a powder bed type system. It also elaborates on identifying many of the issues associated with safety, environment, design considerations, software, machine setup, operations, and post-finishing of parts.

It is not as simple as buying a machine, loading a file, and pushing a GO button. AM involves hazardous operations requiring a dedicated facility, highly trained workers, strict adherence to procedures, a steep learning curve, proper maintenance and fairly high operational costs for metals, gases, and other equipment. That is why jumping right into your own AM system capability requires careful consideration and why using a service provider at the onset may be your best option.

To begin we take you through a number of steps just to give you an idea of what is involved setting up the AM process. The actual procedures will depend on your machine type, material, and parts to be built but they all have similar steps. AM system manufacturers continue to make great strides in identification and mitigation of hazards and streamlining the process and procedures, taking the operator out of the loop, and simplifying process setup.

Putting safety first identifies the need to check and assure all required facility, machine, and personal safeguards. AM systems create hazards beyond those of your typical machine shop and require an extra level of knowledge, planning, and control to keep operations safe, such as those with the use of lasers, powders, and pressurized gas. Posting restrictions for room access or verifying the availability of personal protective equipment such as respirators, eyewear, and protective clothing are a few examples. Are the inert gas bottles in place, of correct type and of sufficient volume and pressure to complete this build? Has the feed stock been properly stored and is it ready to transport and load with the proper carts, dollies and other safety devices? Are anti-static devices such as standing pads in place and properly connected? Are all system maintenance logs up to date with old filters properly disposed of and new filters installed?

Due to the hazards of handling and loading powder, proper procedures must be established and followed. The manufacturer will provide guidelines and training, but if you are working with nonstandard materials you may need to bring in experts to assure a safe process, such as ES&H (Environment, Safety and Health) experts to assure you understand and comply with the proper safety procedures. Other experts such as laser, ventilation or electrical safety officers may need to be consulted. The build chamber setup will include assuring a clean chamber and if other metal powder has been previously used, a full chamber cleaning will be required. If the same powder type is used cleaning may simply require the wiping of condensed metal vapors from chamber surfaces, doors and other hardware. Cleaning of vapor deposits from the laser optics is required after every use. Electron beam or arc heat sources will have their own checks and procedures. Some systems provide diagnostics to measure beam power or beam profile characteristics to assure a clean optical path. Electron beam gun diagnostics can be used in the same manner, including a check to assure the electron filament life is sufficient for the build.

The powder spreading and recoating mechanism can be checked for proper alignment and orientation of the build plate to the recoating blade. The thickness and condition of the plate itself will need to be checked to assure it is flat and free from debris, warpage, or damage from previous builds. The loading of feedstock powder needs a final check to assure previously used powder is properly sieved, baked out, and stored with all traceability and usage records up to date. After the chamber is closed and sealed, preheating to working temperatures and inert gas purging may begin. System interlocks to room oxygen sensor levels or door closures may be checked to assure they are functioning correctly.

To get the most out of building your first part it can pay to consider what type of process monitoring and data gathering is offered by your system or what type of third-party monitoring devices may be installed to gain some insight into the build process.

10.3.1 Monitoring of Process Quality

The fabrication of test coupons and prototypes provides an opportunity to characterize the process as it is being developed. Establishing the signature or baseline of a process, operating normally and under control, is key to verifying normal and off normal operating conditions, especially in small lot sizes where statistical quality control methods cannot be used. Various methods may be used to monitor the system to define an acceptable range of process performance. Capturing the signature of an off normal or a loss of process control is just as important. Blindly producing test samples without process monitoring is a missed opportunity. Parts produced for form, fit, marketing or proof of concept may not need the level of rigor associated with functional prototype or pre-production development units, but if the information is available it should be collected and archived.

As stated earlier in the book, beam characterization is an established way to verify heat source conditions. Characterizing the heat source may be performed before the actual build cycle and is one example of testing subsystem performance prior to the actual build. All of these high energy heat sources degrade in performance over time requiring monitoring, maintenance, and repair. Heat source monitoring may involve beam profiling for laser or examination of the EB gun and cathode condition or inspection of wire electrode contacts for arc systems. Other process disturbances can occur during the build cycle and may be monitored and recorded to document the process. Power supply performance monitoring is often provided by the manufacturer and may be used to construct data sets for each coupon or part built. The costs of AM printing may justify the use of third-party data collection systems for critical applications. The duty cycle of these systems running for hours or days places a premium upon the assurance that the system is performing as it should throughout the entire build cycle. In-process monitoring may be used to infer part quality, but measurements made on the actual part while it is being built can provide a greater level of assurance.

10.3.2   In-Process Part Quality Monitoring

Monitoring part quality during the build cycle may provide additional assurance that the representative regions of the part were deposited as-planned without deviations or disturbances leading to discontinuities or defects. Methods such as these are particularly attractive due to the complexities of applying conventional nondestructive techniques to AM metal parts. Not all AM vendors offer in-process quality assurance (QA), as some of the systems on the market are still at the demonstration level of functionality.

There are strong economic benefits to interrupting a long duration build cycle if the part is deemed to be defective during the build. The difficulty lies in the observability and characterization of part quality during the build cycle. Part visibility in powder bed systems is limited to the last layer fused and may not be representative of that location within the final part, as penetration into previous layers may be required to remelt and heal surface discontinuities and apparent defects. Full deposit density or quality may only be achieved three layers down from where you can observe the process. In addition, melting and penetration from subsequent deposition passes may create subsurface bulk defects such as voids, or undesirable surface conditions as those in the downward facing regions of steep overhangs unobservable by the inspection of the last layer of deposit.

Video monitoring, thermography, in-process measurement and collection of processing data are ways to create a baseline record of the part as it is being built for evaluation either in real time or after the cycle is complete. PBF processes can record this data for each layer of the build, potentially allowing interrogation of every layer surface as deposited during the build cycle. Again, as was mentioned above, does this view of the process tell you everything you need to know about the quality of the process and final product?

Archival storage of the process and part quality record may allow subsequent analysis of a part that has failed in service. Database storage of process development data along with the original CAD designs may be accumulated and archived following the part from cradle to grave, beyond manufacture and throughout its service life. As will be mentioned in more detail later, knowledge of what you are looking for and where to look is critical for detection. Analysis of what was detected is critical for comparison with standards and ultimately for acceptance. Knowing the material and types of defects or flaws you are looking for goes a long way in defining the needs and usefulness of process monitoring methods.

Observability of DED processes is greater than for PBF processes allowing a more detailed interrogation of the part as it is being built. The ability to observe the melt pool region and the potential of observing the entire part, such as to monitor distortion during the build, exists without being obscured by a powder bed. Difficulties due to rougher surface quality may affect the ability to measure and potentially interrupt, interact with or control the process. Other quality issues, such as distortion, may not be fully characterized until the build has completed and the part cooled and removed from the work environment. Having said all that, in-process part quality assurance is an active area of research and development. QA systems for AM are on the market and vendors are assisting customers to understand how to apply them and use them to the greatest advantage. Critical applications of AM parts will require quality systems certified to accepted standards. Hybrid systems integrating CNC and AM may also feature on-machine gauging, contact measurements, or 3D scanning. Depending on your final requirements, an in-process part quality assurance system may already exist to meet your needs.

Integration of in-process part monitoring data with modeling and simulation results is also an area of study. Data driven models, not based on first principal physics, have proven useful in other applications by fitting low order functions to map input to output conditions with low computational overhead and are being proposed for use in AM. The rapid evolution of computer based FEA modeling and prediction may also help predict and control part accuracy and distortion using pre-build simulations of the process. Simulations may be used to accommodate part distortion by geometric compensation of the part model prior to the part build. Many AM systems already offer real-time melt pool sensing and control, but further work is needed to optimize those capabilities for gathering of sufficient data for in-process quality assurance.

10.4   Repair or Restart Procedures

What happens when a problem occurs during a build and the sequence needs to be stopped, or if a defect is identified during a post-build inspection? Can the process be restarted or can the component be repaired? Long duration build times may be considered a vulnerability of the AM process. Processes may be interrupted for any number of reasons, including failure or interruption of any of the primary process parameters, machine subsystem failures, such as inert gas interruption, or a general facility failure, such as a power outage. Having a repair procedure in place or agreeing on a failure recovery plan with your AM service provider may save everyone time and money while still producing an acceptable result.

Restarting a process may depend upon the conditions of process termination. If the process is terminated in a powder bed system due to a loss of power or inert gas supply, a restart may simply require reestablishment of a preheat or inert supply gas conditions. Restarts of a DED process may require preheat, leveling or a surface smoothing passes without the addition of filler to reestablish acceptable deposition conditions. A failure in the material feed conditions in DED, such as a clogged nozzle, may result in a termination crater, a region requiring smoothing of the deposit or accommodating an under fill defect by modifying the restart sequence.

A key condition of a restart would be knowing exactly how and where the process terminated during the build sequence and having the ability to begin again while applying any needed pre-start conditions. Preconditions for restart may require a beam-on dwell time before motion or a dwell in material feed (or skipping the first recoat layer). This would require the system to store its location and all current build parameters in a nonvolatile memory location such as a circular buffer, allowing access, modification, and restart. Modification could involve dwell time, power increases and focus conditions or any of the other primary variables.

A repair scenario may include the repair of a defect detected during post processing or an inspection operation. If a surface or feature needs to be rebuilt, a reengineered build sequence may be needed using all or part of the original STL or CNC file and modifying it to allow the redeposit of an entirely new section. This may entail machining off a damaged or defective region, scanning the part, comparing it to the original model to define the missing region, creating a new build sequence, a new holding fixture and finally depositing the missing region. Large or complicated parts of high dollar materials may justify extensive rework efforts.

Laser clad repair of complex shapes such as turbine blade tips, burner tips or sealing surfaces are commonly applied in industry so a look to these type of procedures may provide insight into the reclamation of parts from failed AM builds.[9]

In the next chapter, we review post-build procedures. If you are having a service provider build the part you should know what your post-processing requirements are, what the vendor can provide and how much they will charge. You may also consider if you want to post process it yourself or have a third party do the job.

10.5   Key Take Away Points

- The many parameters or degrees of freedom offered by AM processes offer a powerful opportunity to build parts beyond the capability of conventional methods. This freedom has its drawbacks related to the complexity of process selection and optimization.
- Conventional processing methods have taken decades if not centuries to understand, optimize and refine. While AM metal process have existed for decades the bulk of application and process development has been more recent as the numbers of commercially available AM metal machines have increased and become available to industry.
- The majority of parts produced by AM metal processes require some form of post processing. Complex AM designs or advanced materials can produce unique challenges to conventional post processing operations.
- Real-time monitoring or process performance and building the part are challenged by limitations to observe and accurately measure the process conditions. Data gather during processing may serve to augment challenges to existing inspection methods.

---

[9]EU Cordis, Fraunhofer, SLM repair, http://cordis.europa.eu/news/rcn/32171_en.html, (accessed March 22, 2015).

Chapter 11
Building, Post-Processing, and Inspecting

Abstract The build cycle of an AM part can be broken down into prebuild operations, the actual build, post-processing, and inspection. This chapter provides the reader with a typical scope and sequence of operations and considerations associated with these operations. This knowledge is useful for those users considering the purchase and establishment of an AM system capability as well as those utilizing service providers. Knowledge of post-processing and finishing operations is critical to the design process and plays a critical role in subsequent design or process improvements. Operations such as powder recycling, heat treatment, hot isostatic pressing, machining, and surface finishing are discussed as well as the application of nondestructive and destructive evaluation and defect detection as applied to AM metal parts. Ongoing efforts within industry to draft standards and certify AM produced part are described.

11.1 Building the Part

The start sequence is critical to establish and verify proper process performance before transitioning to fully automated operation. Once you have assured all subsystems are functioning together properly and have initiated any process monitoring, you are good to go. Having an AM operator on call to attend or to a stop the operation is needed to assure proper termination, removing the part and prepare the system for the next build. Certain AM system designs reduce the downtime between building one component and the next by utilizing dual or modular powder bed configurations, allowing part or powder module removal of a completed build simultaneous with starting the next.

The successful completion of a build cycle often includes a cooling interval particularly for EB, DED and arc-based systems. PBF-EB, PBF-L, and DED-L are more efficient in melting and experience less heat buildup due to the high-energy heat source. However, PBF-EB uses a powder bed preheat of up to $\sim 700\ ^{\circ}C$ and is performed in a vacuum chamber that slows heat transfer and cooling after the build cycle due to the preheating of the powder bed, combined with the additional heat

© Springer International Publishing AG 2017
J.O. Milewski, *Additive Manufacturing of Metals*, Springer Series
in Materials Science 258, DOI 10.1007/978-3-319-58205-4_11

input of the electron beam. An inert gas purge of the chamber may be required to assist in heat transfer and cooling until the part is below a reactive temperature and allowed to cool in air to a temperature that allows effective handling, powder removal and post-processing.

In the case of PBF-L systems, shorter cool down interval is required as powder bed heating is less than with PBF-EB. After cool down, removal of the powder will include special powder handling procedures to allow removal of the build platform, the part, and support structures.

Powder Reclamation

Powders optimized for AM use are often expensive therefore reclaiming and reusing them is critical to the process. The claims that used powder can be re-sieved and reused indefinitely are disputed. Changes to powders may include losses due to remelting, vaporization, oxidation, and moisture pick resulting from normal processing, or improper handling or storage. These changes may in many cases be insignificant with slight charges to particle size distribution but will also be dependent upon the alloys used and types of powder. This is an active area of research (Ardila 2014), but for now it is best to follow the vendor's recommended procedures. Powder vendors are beginning to fill the need by offering powder characterization services and tracking software to assist customers in maintaining their AM powder inventory.[1]

Removal of the support structures and part from the build plate may require conventional machining, sawing or EDM (electrode discharge machining) operations as well as finishing of the build plate and measurement to assume a minimum build plate thickness and flatness specs are met. Post-build procedures may include cleaning of the chamber or post-processing the powder by sieving to allow reuse and recycle. In some cases, a part may be left connected to the build platform as a support fixture for subsequent finishing operations such as machining or inspection. In other cases, the build plate may become integral to the final part following through all post-process heat treatment and machining operations.

11.2 Post-Processing and Finishing

What sorts of post-processing operations (Fig. 11.1) are required? Can you perform these operations or will you send the parts to a service provider to complete the post-build finishing operation?

Full functionality of a deposited metal part will most often require post-processing and finishing to achieve the desired dimensions and properties. Post-processing operations for metals may require heat treatments, machining

---

[1]LPW Web site describes the roll out of their POWDERSOLVE service, software and online database, http://3dprint.com/53072/lpw-powdersolve-metal-powders/, (accessed March 26, 2015).

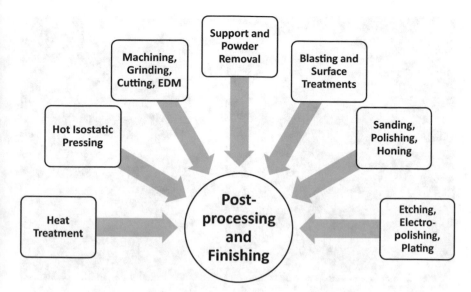

Fig. 11.1 Post-Processing and finishing operations

operations, and access to resources that require specialized precision equipment, expert knowledge, and operations beyond simple media blasting, sanding, or coating. Knowledge of these post-processing operations, where and how to access these resources, will help you better choose the process and materials to make the best up-front design decisions. Knowing how and when to apply these operations is critical to efficiently achieve the desired full performance of your metal part.

One of the first operations after completion of a build cycle and cooling is to remove the part from the powder bed, build platen, or fixtures. As introduced above, this will include powder removal, recovery, and physically removing support structures or fixtures, emptying powder volumes internal to the part through drain holes and clearing internal passages. Figure 11.2a, b shows a support structure and its removal for a nozzle component. Recovering the powder and recycling it back into the process for subsequent builds may require different procedures depending upon the powder type and build conditions. Recommended practices for tracking and mixing new virgin powder with sieved and reused powders will vary depending on applications, standards, and certifications still in development.

Finishing may include media blasting, peening, sanding, abrasive slurry honing, or grinding to smooth surface features and allow visual inspection. Washing may be used to help remove powders from internal features. In cases such as medical applications, sterilization may be specified. Coating and painting, as used in plastic prototype finishing, may improve surface finish or appearance. Alternatively, smoothing or finishing may employ slurry polishing, electro-etch, electro-polishing, or plating operations. These operations may require specialized equipment and processes provided by a dedicated service provider.

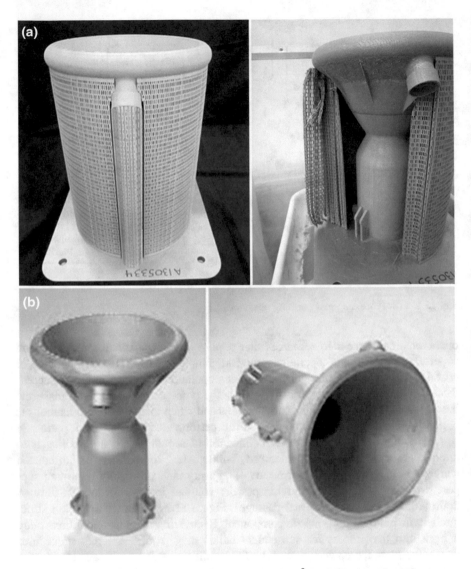

Fig. 11.2 a As-built nozzle structure with support structure.[2] **b** Finished nozzle with supports removed[3]

Partially fused powder particles may become dislodged and affect in-service part performance. Machining, grinding, polishing, or coating are all candidates to modify as-deposited powder, but the more post-processing required by your design, the more you diverge from having a straight-out-of-the-machine functional object

---

[2]Figure courtesy of Lawrence Livermore National Laboratory, reproduced with permission.
[3]Figure courtesy of Lawrence Livermore National Laboratory, reproduced with permission.

and the more you reduce the benefits of AM fabrication, when compared to conventional processing. Hybrid systems combining improvements in laser optics and beam delivery may extend capabilities to include automated finishing or inspection into the build space.

External supports or base features may need to be removed as part of the post-build finishing operations. The design of support structures and optimization to allow ease of removal, is incorporated into the build cycle using software, while the removal of supports is typically performed by sawing, cutting, machining, grinding, EDM, or other mechanical means. Knowledge of what post-processing methods are best for a specific material and design is critical to optimizing the AM process for a specific part.

Heat treatment is often required to achieve the engineering properties of the AM metal. Heat treatments may require 2–4 h in an inert or vacuum furnace at temperatures ranging from 650 to 1150 °C. These treatments may be required to improve or meet the desired strength, hardness and ductility, fatigue, or bulk properties. As stated earlier in the book, heat treatments, such as annealing, homogenization, solutionizing, or recrystallization, may be needed to achieve uniform bulk properties or achieve the desired microstructure. The layer-by-layer deposition can result in directionally dependent properties and can vary with respect to the orientation of the part within the build chamber. HT furnaces used to treat metal parts may need to operate at high temperatures and use inert atmospheres, such as argon or vacuum, when processing certain materials. Relief of residual stresses present in an AM part may also require HT to assure dimensional stability. All these operations require specialized HT equipment and may take hours to complete. Research is ongoing to understand and define heat treatment conditions for AM deposited materials. In one example (Mantrala et al. 2015) research indicated differing heat treatments of the same material were required to optimize either hardness or wear resistance.

As mentioned earlier, hot isostatic pressing (HIP) is a process that uses high temperatures and high gas overpressures to heat a part to a temperature below melting and at pressures of 100's of MPa, and temperatures in the range of 900–1000 °C for 2–4 h to help close and fuse internal pores, voids, and defects. HIP can also provide heat treatment benefits by optimizing the temperature and pressure cycles to improve mechanical properties such as strength, elongation, ductility, and to improve the structural integrity of the component. The equipment is large and costly and may require a specialty service provider. HIP pressure chambers typically are limited in size ranging from ~ 75 mm to 2 meters, limiting the size of an AM part to be consolidated, although custom HIP system designs and services are available in industry.[4]

---

[4]As an example, Avure is now technically capable of producing HIP pressure vessels up to 3 m (118") in diameter. Standard pressures from 1035 to 3100 bar (15,000–45,000 psi) are available, other pressure vessel designs can be supplied to meet individual processing requirements, featuring a variety of styles, hot zone materials, rapid cooling technology, and advanced temperature measurement techniques, http://industry.avure.com/products/hot-isostatic-presses, (accessed March 26, 2015).

High precision parts will often require subtractive post-processing operations, such as machining, grinding, or drilling, to achieve the dimensional tolerances and surface finishes of the final functional shape. Specialty operations such as plunge EDM or polishing may be needed to achieve the final surface contours of molds, punches, and dies. The tradeoffs between the accuracy, surface finish or microstructure of the as-deposited part and the desired final finish of the part entail a complex set of decisions. Regardless of the process used to fabricate the part, either by PBF or DED, laser, electron beam, arc, powder, or wire, a careful evaluation of the final requirements will contribute to the decisions made during process, material and procedure selection.

Creating machining blanks may relax the dimensional requirements of the deposit. In cases such as in DED, deposition rates may be more important than accuracy. If the as-deposited near net shape requires machining to achieve the final dimensions, surface finish or distortion may be less important than optimizing the build rate. As an example discussed earlier in the book, DED-EB may be used to create very large objects to be subsequently machined to final dimension. The cost savings result from not having to machine very large billets resulting in the creation of a significant amount of wasted material. In cases such as these, the degree of stair stepping, distortion, and relatively crude deposition to the near net shape is less important because the final shape and tolerances will be achieved by machining.

The hybrid combination of DED and CNC machining holds promise to incorporate DED using lasers and arc-based systems into multiaxis machining centers. Again, speed may take precedence over accuracy if all the AM deposit will be machined to final dimension.

11.3   Bulk Deposit Defects

Figure 11.3 shows some of the common defects present in AM metal parts.

Deposit quality is judged by the properties achieved and the ability to meet the design requirements. If you are AM building a component for critical service, such as an aerospace part, part failures can have grave consequences. Therefore, the appropriate level of inspection and quality control becomes important. AM design and deposit requirements are set by engineers with knowledge of what size, number, and type of flaws are allowable and the confidence the process is under control and able to meet these requirements. Qualified operators, procedures, formal inspection, and records are as important as the parts themselves.

The functional requirements of components are not limited to structural and aesthetic needs. As an example, if high pressure, high temperature, impulse loading, or highly corrosive environments will be encountered by the part, the properties should be appropriate for the service conditions. Knowledge of AM defects that can affect the material properties or lead to failure is critical. Knowing how defects occur and how to detect them, using destructive and nondestructive inspection methods, will help you select the most appropriate inspection methods for these specific types of defects.

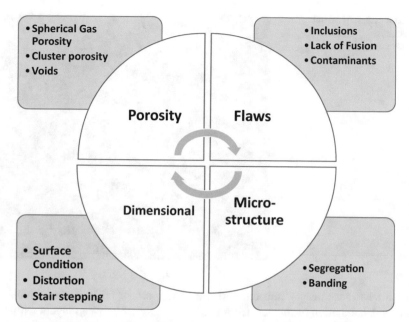

Fig. 11.3 AM metal flaw and defect types

Due to the complex nature of the AM processes, the bulk material character or morphology of the deposit can vary widely from one process to the next. It will be different for different materials and may vary widely from one location in the part to another. This may be due in part to changes in the many variables and conditions during deposition. As an example, changes in energy input may change the thermal conditions and the ability to accurately deposit certain features. Defects not within the bulk material, but within the character of the part, such as distortion, may also occur.

Flaws are undesirable conditions within the deposit, while *defects* are flaws that exceed an acceptance criteria. The detection of defects can trigger actions such as rework, repair, or part rejection. AM metal defects may include lack of fusion, porosity, voids, cracking, oxidation, discoloration, distortion, irregular surface profile, or surface stair stepping. Later we discuss common AM inspection techniques including visual inspection, dimensional inspection, dye penetrant, leak testing, radiographic, computed tomography, and proof testing.

Defects typical to fused or sintered metal processing are often present to one degree or another in 3D fused metal parts. Given the intrinsic need for accuracy in PBF parts, the molten pool size is typically smaller and the cooling rates are typically faster than common welded structures, resulting in defect morphology more typical to laser or EB welds rather than large arc welds or castings. DED processes with large molten pools and high deposition rates produce defect common to welded structures.

Lack of fusion is one of the more common defects where the bead or deposit does not fully melt and fuse into adjacent tracks or into the substrate. *Lack of penetration*

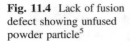

Fig. 11.4 Lack of fusion
defect showing unfused
powder particle[5]

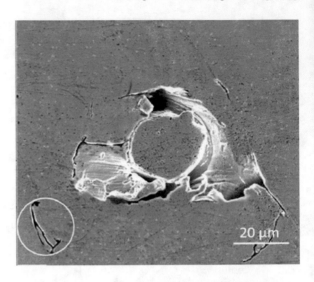

is a failure to fuse deeply into the part, build support, or build platen, which may
result in delamination of the part from the build platen or between layers within the
deposited part. Lack of fusion may be visually detected but it can also go unnoticed
until a part fails in testing or service. Radiographic inspection may or may not
detect lack of fusion defects as poorly bonding locations may not be detected.
A *cold lap* is a term sometimes used to describe a lack of fusion defect in which
coalescence of material is prevented by an oxide layer or thin film. A poorly fused
powder particle boundary, as shown in Fig. 11.4, may crack or fail in service.
A failure during destructive testing, functional testing, or while in service should be
inspected for lack of fusion and the AM build schedule modified accordingly. In
some cases, flaws leading to fracture initiation points fracture initiation points may
only be observed under high magnification using scanning electron microscopy
(SEM) and a metallurgical inspection technique known as *fractography*. We will
discuss these later under failure analysis.

   Slumping is a dimensional defect typically created in regions built with insufficient support structure. Slumping defects may be associated with small design
features, thin walls, downward facing surfaces (also referred to as down skin surfaces), or overhangs. If the molten pool is too large for the position, gravity can sag
or slump the molten pool and create this type of defect. Rounding or loss of edge
quality may occur when *surface tension* draws a molten pool into a rounded shape
resulting in the loss of dimensional fidelity of sharp edges or corner. If the heat
source power is too high and deposit or traverse speed is too fast, surface tension

---

[5]*Source* Hengfeng Gu, Haijun Gong, Deepankar Pal, Khalid Raf, Thomas Starr, Brent Stucker,
"Influences of Energy Density on Porosity and Microstructure of Selective Laser Melted 17-4PH
Stainless Steel", D.L. Bourell, et al., eds., Austin TX (2013) pp. 474–489. Reproduced with
permission.

can draw the weld bead up into a *ropey* shape leaving a lack of fusion along either side of the deposit. These sorts of instabilities may also lead to *spatter* and droplets of melt or partially melted powder particles from the melt region, further disrupting the deposition quality. Proper parameter selection such as control of beam power or speed can help to fuse and flatten the deposit bead. *Undercutting* is a flaw typical to fusion processing that can occur in regions where gravity or surface tension draws molten metal away from the edge of the pool surface leaving an under filled region that may act as a *stress riser* or crack initiation point.

Shrinkage and distortion can occur as a result of localized melting and solidification. Shrinkage-induced stresses can build up creating warping and distortion and in some cases reach levels where lack of fusion and delamination can result as shown in Fig. 11.5. Proper design, material and parameter selection and process development can reduce or eliminate the risk of these types of defects.

Irregular surface condition can indicate a poorly designed, supported, or oriented part. A poorly developed or controlled process may result in *stair stepping, balling,* lack of fusion, surface breaking delamination, undercutting, holes, porosity, or voids. Excessive spatter of balls fused powder particle ejected from the melt region may be fused to either side of the deposit, indicating improper beam power, focal conditions, contaminated powder, or filler wire. Irregular surface condition can reduce part strength by creating stress concentration locations that may fail or fatigue during service. Notches, voids, or undercuts can concentrate stresses and initiate cracks. Cosmetic and surface requirements not met will require post-processing to either remove or finish the top surface.

Porosity is a common defect evolved in melted and fused material, often spherical or oblong, resulting from gases, such as hydrogen entrapped within the molten pool and evolved and released as bubbles upon cooling and solidification. Another source of porosity is the melting of an unfused region or void in which the gas within the void forms a bubble within the liquid and entrapped within the solidified metal. Keyhole collapse as in laser or EB melting can entrap gas within the melt pool during solidification.

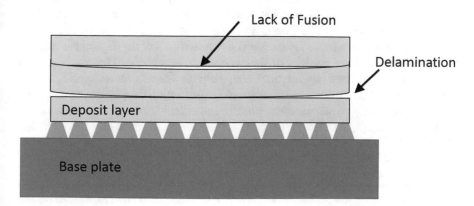

Fig. 11.5 Lack of fusion or delamination defects

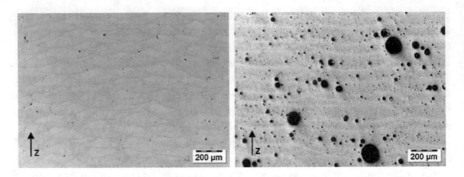

Fig. 11.6 Hydrogen porosity within SLM deposit of AlSi10Mg[6]

It is important to differentiate between gas porosity and other forms of voids such as lack of fusion voids as different mechanisms lead to their formation and need to be considered separately to assure proper control. Sources of these gases may be moisture or contamination of the inert gas atmosphere, build chamber, powder, or filler supply. Hydrogen as a gas is readily absorbed into molten metal and can be rejected from the melt during solidification and trapped as bubbles or pores. Hydrogen contamination of the feed stock or the process before or during the build cycle can result from a number of sources such as improper quality control of the wire drawing process or improper storage. Porosity is more of a problem for some materials than others, such as aluminum and those more susceptible to gas absorption during powder or melt processing. Strict handling, storage, and processing procedures are needed to control sources of porosity. Argon may be trapped into powder as a result of the gas atomization process and may be another source of micro-porosity. Entrapped gases such as these may coalesce during melting and result in the growth of porosity during solidification.

Pores can reduce the cross-sectional area of the deposit thereby reducing strength, although spherical porosity is not as critical to loss of strength as angular crack initiating defects that have a greater tendency to propagate under loads. Surface breaking porosity and voids can hold water or moisture and exacerbate corrosion and staining. The AM designer or fabricator needs to consider all possible defect formation scenarios to assure that the requirements of the deposit are met and to decide what level of porosity or flaw content is acceptable. The distribution of porosity within a fused region may also help to identify its origin. Spherical porosity may indicate a contamination source either within the powder, atmosphere, or resulting from contamination during storage, handling, or processing. Figure 11.6 shows two levels of porosity in an SLM deposit of AlSi10 Mg achieved by drying the powder and modifying the process parameters.

---

[6]*Source* C. Weingarten et al., J. Matl. Proc. Tech, 221, (Elsevier 2015) 112–120, http://dx.doi.org/10.1016/j.jmatprotec.2015.02.013. Reproduced with permission.

Voids not evolved from gas rejection or gas entrapment upon solidification may take many different forms and can be the result of many different processing conditions. Lack of fusion voids common to the AM PBF processes include those associated with the powder packing density, or the spaces in between powder particles, when spread in layers and inadequately fused, resulting in a poorly fused deposit. The microstructure of sintered powders, resulting in less than fully dense deposits, will display unfused regions or voids requiring remelting through subsequent layers of deposit or additional HIP post-processing. These regions may contain unfused powder particles or inadequately fused particles due to insufficient liquid phases while sintering, or in other cases insufficient melting or mixing within the molten pool. As an example, a decrease in beam energy density may result from an optical component that is misaligned or needs cleaning, resulting in a lack of fusion defect. Conversely, increased energy density resulting in a deviation of scan speed, e.g., during a change in scan direction, may create a localized vaporization event or an unplanned transition to a keyhole mode of melting resulting in entrapped porosity due to keyhole collapse.

Localized *inclusion* defects may result from contamination contained within the powder or filler wire which may vaporize during processing with sufficient force to eject molten material from the melt pool leaving a hole or void that may not fill during the fusion of subsequent layers. As an example, erosion or damage to the recoating blade powder spreader or rake may leave particles of foreign material within the build material supply.

Constraint is a term often used when discussing the formation and avoidance of cracks associated with metal fusion. As discussed earlier in the book, metal expands or contracts when heated or cooled. As a result it may either grow, shrink, warp, or bend. If heated or cooled and constrained from moving, such as by clamping, the use of support structures, or because mechanical constraints within the design itself, stresses will build up. These stresses are mechanical forces that can either be locked up and reside in distorted crystalline structure, or be relieved by distortion, cracking, or tearing. The degree to which movement is allowed or prevented can be referred to as constraint. *Residual stresses* are difficult to measure and although not often classified as flaws, may lead to distortion or cracking in service. They are often controlled through the part design, processing condition selection or post-build heat treatments.

Cracking can result from a wide range of thermal, mechanical, and metallurgical conditions. Figure 11.7 shows cracks in a welded reactor vessel as detected by dye penetrant testing. AM components can be susceptible to the same types of cracking mechanisms present in welded or weld clad structures. A few of the more common types of cracks are *crater cracks, hot cracks, hot tearing, and cold cracking*. Some materials are much more sensitive than others to the development of cracks as a result of AM metal processing. Crater cracking refers to cracking that may occur at the termination of an AM deposition path within the last to solidify material. Hot tearing may occur in regions directly adjacent to the fusion boundary, at temperatures below melting, in the metal softened by heating. Hot cracks may occur directly upon cooling near the solidification boundary, while cold cracks (also known as delayed cracking) may occur hours or days after cooling. The metallurgical reasons for these types of

Fig. 11.7 Cracks in a welded
reactor vessel as detected by
dye penetrant.[7]

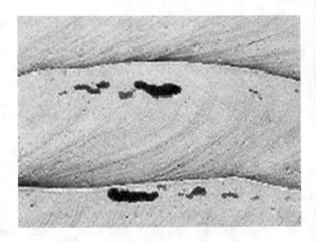

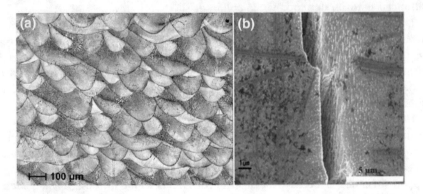

Fig. 11.8 a Microcracking of DMLS Inconel 718[8] **b** Microcrack opening in Inconel 625 shown
using scanning electron microscopy[9]

defects range widely and are beyond the scope of this book, but the reader should know
these conditions exist and consult a professional metallurgist when working with new
alloys or with potentially crack sensitive materials. AM vendors have developed strict
control of materials and parameter sets to avoid many of these types of problems, but if
you are developing your own parameters for materials that are crack sensitive, be ready
to address these problems. Cracks can occur at the termination or end point of an AM

---

[7]Courtesy of Enspec Technology, reproduced with permission.

[8]*Source* Ben Fulcher, David K. Leigh, "METALS ADDITIVE MANUFACTURING•
DEVELOPMENT AT HARVEST TECHNOLOGIES," SFF Symposium Proceedings, D.L.
Bourell, et al., eds., Austin TX (2014), pp. 408–423. Reproduced with permission.

[9]*Source* Li Shuai, Qingsong Wei, Yusheng Shi, Jie Zhang, Li Wei, "Micro-Crack Formation and
Controlling of Inconel625 Parts," SFF Symposium Proceedings, D.L. Bourell, et al., eds.,
Austin TX (2016), pp. 520–529. Reproduced with permission.

melt track or weld. The last part of the bead to solidify results in a depression and shrinkage and may result in cracking. A crater depression with cracking can concentrate stress induced by in-service loads and propagate the cracks possibly leading to failure. Properly developed procedures, such as those supplied by vendors, and proper selection, and control of materials will avoid these types of defects.

Hot cracks and tears form as the molten pool solidifies. They are often formed by shrinkage stresses induced during cooling due to a combination of chemical, metallurgical, and mechanical conditions of the solidification region, partially melted or low ductility heat affected region. Highly constrained part locations, unable to flex and distort due to thermal expansion and contraction, can literally pull apart or tear when the metal is still hot and weak during cooling. Recall the discussion early in the book regarding thermal softening of metals. As with porosity, crack location and character can indicate the mechanism of formation and assist in detection, prevention, and control. As with fusion welding, many types of cracks may be associated with AM deposits. Figure 11.8a, b shows microcracking in direct metal laser sintered Inconel 718 produced under experimental conditions. Micrometer scale cracks can occur along grain boundaries and across grains and back fill with the last to solidify material making them hard to detect. Proper parameter selection, preheating conditions, or HIP processing can reduce or eliminate the formation of cracks and bond defects in crack sensitive materials. Undetected microcracks within components under service conditions such as cyclical loading, may eventually lead to crack propagation and in-service failure.

Hot cracking is commonly encountered in aluminum alloys but can often be avoided by proper selection of part geometry, metal, and parameter selection. Stress relief features may be designed into the part or support structure to relieve stresses near crack susceptible regions. Cold cracking, as the name implies, refers to cracks that can form hours, days, or much later in the service life of a component with the potential for catastrophic failure.

There are many more types and reasons for cracks, but suffice it to say that cracking in AM deposits may be one of the more serious defects, as a small crack initiation site can result in catastrophic failure in service. Earlier we discussed some design techniques used to avoid distortion and cracking. Later in the book, we discuss some ways to detect cracks and avoid cracking in AM parts as the recommended practices being developed may deviate from those of conventional processing, such as those typically used for welded fabrication.

Oxidation and discoloration can occur when the molten pool or surrounding hot metal is improperly shielded from air and atmospheric conditions. Chemical reactions take place at different temperatures forming compounds of different colors on the part surface. Changes in chemistry may also occur below the surface. Changes in the build schedule, process disturbances, interruptions, or other sources of contamination can affect the surface chemistry. Regular maintenance, source material control and proper setup may help reduce sources of contamination and discoloration. Routine use of equipment checks to look for problems, such as loose fittings or problems with the gas supply, is also warranted.

11.4   Dimensional Accuracy, Shrinkage, and Distortion

The accuracy of a desired part is largely determined by its design, processing conditions and material. As mentioned earlier, the designer must take into account the thermal and mechanical characteristics of the material being processed. The build sequence or schedule must appropriately specify deposition conditions to account for localized thermal expansion and contraction during the build cycle. *Contouring* conditions, or those build parameters used along the perimeter of each AM layer, must be tailored to achieve the desired accuracy and surface smoothness. Orientation of the part within the build chamber, to minimize stair stepping and layer effects, is needed as well as the appropriate design of support structures used to assist in the deposition of overhang features. Spring back of a part when removed from a base platen or support structures may need to be accounted for in the design. When building a near net shaped component that will require subsequent machining, a material or machining allowance is required. As an example, the material allowance enables the machining process to remove enough of the as-deposited surface to reach sound metal and achieve the required surface condition. For parts requiring high-dimensional accuracy, a machined datum or inspection surface is required to define the location of other inspection features. Another way to describe this is "being able to find the part within the near net shape". Heat treatments such as stress relief, annealing, or HIP processing may induce additional dimensional changes which may require revising the original design, as dimensional changes resulting from heat treatments may be unpredictable during the first design iteration. In the future, computer simulation and prediction tools, such as FEA modeling, may be developed to assist in accurate shrinkage prediction for large complex parts. But for now, trial, error, and experience may be the only option. When it comes to bulk material or part defects, knowing what you are looking for and how defects are formed is key to detection and prevention.

11.5   Inspection, Quality, and Testing of AM Metal Parts

Additive manufacturing processes used for metals present some unique issues associated with inspection, quality assurance, and testing. In the previous section of this book, we discussed the occurrence and type of flaws typical to AM metal parts. As any metal fabricator knows, detecting flaws and defects in a weld is difficult enough, but when the part is entirely made up of welded filler material the potential for those types of defects may be present in any location within the part. The refined nature of the buildup, due to thin deposition layers and fast cooling rates, will lead to defect morphology and geometric conditions that challenge even the best inspection methods currently in use. In some cases, current inspection methods applied to AM produced parts that are fully heat treated or HIP processed and 100% machined may be appropriate, in other cases alternate means for acceptance may need to be developed. As a general rule, as part complexity increases NDE inspect ability decreases. AM processes can add significant design complexity challenging traditional NDE techniques (Todorov et al. 2014). In this section, we discuss available inspection methods and some of the challenges presented by AM metal fabrication.

11.5.1   Nondestructive Test Methods

Figure 11.9 identifies nondestructive test methods being applied to AM parts. Visual inspection is often employed to identify gross defects such as distortions, surface conditions, and gross anomalies. Camera and image inspection can accurately and rapidly capture and classify part features and compare them with a standard definition or part model. Cameras may also utilize magnified views of regions to verify conditions against a standard or population of parts. Borescope inspection may be incorporated to verify clearances or deposit conditions within enclosed volumes or passageways. Multiple image cameras combined with software are capable of measuring distances and other characteristics of AM parts. In another example of visual inspection, titanium surface discoloration can indicate contamination of the powder before or during processing and may be cause for part rejection. Supporting this example, the AWS, *Specification for Welding Procedure and Performance Qualification*,[10] specifies color acceptance criteria for class A, B, C critical welds. Challenges to visual inspection presented by AM can include complex shapes and internal feature beyond the line of sight of inspectors or cameras. Unless the part was specified to be built in the same orientation, using the same support structures and build parameters, the occurrence of surface defects, such as stair stepping, may occur in different locations on parts built to the same 3D model.

Dimensional inspection relies on measurement tools for dimensional verification while gauges may offer a faster way to verify a part dimension or extent by using a go-no-go gauge. Various nondestructive methods may be used to assure the clearance of internal passageways and volumes such as by using displaced liquid, pressure–volume–temperature, flow rate, or picnometry. Coordinate measurement machines (CMM) are precision measurement devices employing a touch probe, manipulated using CNC or robotic motion, to measure the location of part features. Software may take these measurements and compare them to locations on the part definition model. On-machine gauging can incorporate an inspection gauge directly into a CNC milling or turning environment. With the advent of Hybrid AM/SM machines, the logical next step is combining CNC with AM and in-process measurements. Methods such as this are currently under development potentially

---

[10]AWS B2.1 M:2014, *Specification for Welding Procedure and Performance Qualification*.

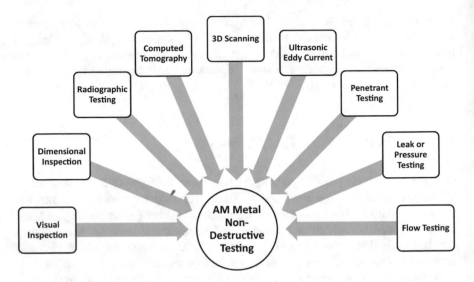

Fig. 11.9 Nondestructive testing techniques for AM

enabling a single machine to produce a fully certified part without the need for subsequent inspection steps.

3D digital scanning refers to a number of techniques used to capture the geometric extent of an object, generate a point cloud of data approximating the surface of an object, and fit a geometric surface or solid description to the point cloud creating a digital model. This point cloud generated solid model may then be compared to the original part model definition. Digital scanners often are structured light or laser based systems that move in relation to the part surface sensing the reflected beam to define the location of data points in 3D space. Surface finish, color, reflectance, smoothness, and lighting can affect the accuracy of the scan and may limit the effectiveness in certain AM applications. As with visual inspection, complex shapes with no line of sight to critical features may eliminate the usefulness in certain AM applications. Deep features, high aspect ratio holes and internal features may not be captured. Commercial scan systems range widely in accuracy and costs as does the supporting software. Software is used to best-fit the point cloud to geometric features and surfaces, but again, the accuracy of the point cloud and the fitting algorithm will affect the final definition of the inspected part.

Radiographic testing (RT) has been used for many years to inspect welds in pipelines, pressure vessels, and a wide range of critical use metal components. Irregular surface conditions or multiple or complex internal features along the X-ray path may obscure features of interest or complicate interpretation of the images. Digital radiography offers a 2D gray scale image that may be enhanced using color or other digital techniques. Microfocus radiography offers greater resolution for a given wall thickness within a narrow field of view.

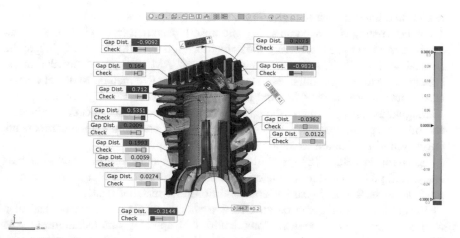

Fig. 11.10 CT CAD to Inspection Model Using Geomagic Control X Software[12]

Computed tomography (CT) relies on a series of images, such as those obtained by X-ray, or magnetic resonance imaging (MRI) taken at specific angles relative to the part and reconstructed by computer software into a 3D data set revealing interior and exterior features. CT inspection models may then be used to compare to the original CAD model enabling part-to-part comparisons (Fig. 11.10). The quality and resolution of the scan and the accuracy of the algorithms used to reconstruct and render the geometry will affect the results. The energy source, detector resolution, and setup will contribute to the measurement accuracy. These methods have been demonstrated to characterize features such as porosity, inclusions, minimum wall thickness, and other internal features. One problem can be that of false positive indications of defects due to the complex microstructure of certain as-built AM parts. While this technology has been successfully demonstrated in the analysis and inspection of AM metal parts, the costs, complexity, and computing requirements of the large data sets generally restrict the use to specialty service providers and for use by customers with high value components.[11]

Ultrasonic testing (UT) uses ultrasonic sound waves to penetrate a metal object, reflect off internal features, and bounce back to a detector to reveal the approximate size and location of these features. It generally relies on a liquid coupling between the probe and the top surface of the object. It may be limited by curved, complex internal and external surfaces or the rough surfaces of AM deposited parts. Post-processing and finishing may be needed prior to the application of UT.

Penetrant testing (PT) is often used to detect cracks or small surface breaking flaws. It uses a liquid penetrant, often a dye, sprayed on the surface to be tested, that

---

[11]More information may found on the Web site of Jesse Garant and Associates, http://jgarantmc. com/, (accessed March 26, 2015).

[12]Courtesy of 3D Systems, reproduced with permission.

is absorbed into the flaw features. The excess surface dye is wiped off and a liquid developer coating is sprayed on to dry and absorb penetrant from within the cracks, crevices, voids, or pores revealing the defect location (refer to Fig. 11.7). While the method may be limited for use on as-deposited AM surfaces of powder based systems, due to surface roughness, it can offer an effective method to detect cracks, lack of fusion or undercuts in weld deposits of DED wire based processes.

Magnaflux testing is often used on large cast components to detect cracks in steel and other magnetic materials. It is often used to find cracks in engine components but may find application as applied to AM repaired parts.

Eddy-current testing (ET) is used to detect surface and subsurface flaws, such as cracks and pits, in conductive materials using electromagnetic induction. It can be sensitive to surface and near surface conditions and material type.

Vacuum leak testing can be an effective way to assure the hermetic seal of a volume or containment vessel. Complications of applying this technique to AM fabricated products include the need for a sealing surface or sealing compound with the rough surface of as-deposited AM parts or the post-machining needed to obtain a flat sealing surface. Sintered, porous product, or partially fused powder surfaces could create a *virtual leak* or hidden pumping volume, reducing the utility of the process.

Flow testing may offer a means to assure clearance of complex passageways without the cost and complexity of a CT system. While many of these NDT procedures may be difficult to apply to as-deposited AM parts, standard test specimens and standardized acceptance criteria will be developed that offer the ability to characterize the bulk deposit using NDT methods combined with DT (destructive test) methods, as described below. A good source for additional information is ASNT, the American Society for Non-Destructive Testing.[13]

11.5.2   Destructive Test Methods

Earlier in the book, within the process development section, destructive testing was mentioned as being used during parameter studies and during the fabrication of test samples. The samples are sectioned and subjected to a variety of metallographic tests to determine the microstructural character, defect morphology and mechanical properties of the deposit as they relate to processing conditions. Small lot production may rely on destructive testing of witness coupons fabricated layer by layer or within the same build cycle of the actual part to infer proper machine function. Figure 11.11 lists categories of destructive test methods being applied to AM parts.

Microstructural analysis will often entail preparing small samples of representative regions of interest within a part and by mounting, polishing and in some cases

---

[13]American Society for Non-Destructive Testing, https://www.asnt.org/, (accessed March 26, 2015).

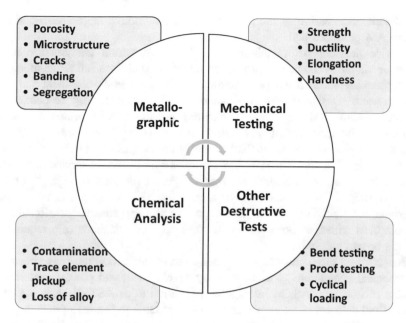

Fig. 11.11 AM Metal Destructive Tests

etching the samples to reveal the structure of grains, phases, inclusions, and defects such as pores, voids, cracks, and other discontinuities. As-polished samples may rely on camera-based microscopic inspection, coupled with image processing, to provide a deposit density estimation, (% voids) or other semiquantitative determination. Hardness values and phase identification can be used to verify time–temperature transformations related to cooling rates and infer property determination confirmed by standard mechanical test procedures. Microstructural analysis coupled with nondestructive testing and proof testing is often all that is needed to qualify a noncritical component or process. A source for procedures used for microstructural analysis can be found in the tech sheets of the Beuhler[14] Web site.

For common engineering materials in commercial shapes, mechanical property data contained within engineering data books such as those provided by Society of Automotive Engineers (SAE), military specifications (MIL), and The American Society of Mechanical Engineers (ASME) is often sufficient to provide the confidence and documentation required by a quality trail. In cases where standard test data is unavailable, a full material and process qualification may need to be performed to assure the material properties and part performance. Vendors of AM machines often supply this information for their proprietary material, produced using their standard parameter sets and using their stated test conditions. In some cases, this may be sufficient to infer the integrity of a functional prototype, although

---

[14]Beuhler Web site support has additional support information regarding metallographic preparation, http://www.buehler.com/, (accessed March 26, 2015).

the material test conditions and properties cited may not be representative of those of any AM produced part.

Service providers of metallurgical and microstructural analysis are readily available via the Web and can provide rapid turnaround for a full range of services to characterize and formally perform analysis for process development and qualification needs. Design definition of standard test specimens may be enhanced to include standard AM deposition and testing conditions with which to compare material lot chemistry variations or machine to machine differences.

Chemical analysis can be applied to verify the integrity and purity of materials and processing conditions used in AM. Samples may be sent to service providers to identify levels of contaminants such as oxygen, hydrogen, nitrogen,[15] or trace elements such as iron, vanadium, or aluminum. This analysis can be useful to determine the pickup of contaminants or loss of alloying constituents. Results may be compared with feed powder or wire chemistry specifications as a required for process or part acceptance.

Standardized mechanical test specimens and testing procedures have been used for technical generations to assist in alloy development and process material and process certification. ASTM International (referred to as ASTM) leads the development and delivery of voluntary consensus standards to improve product quality, enhance health and safety, strengthen market access and trade, and build consumer confidence.[16] American National Standards Institute (ANSI) serves to administer and coordinate the private-sector voluntary standardization system, for creation, dissemination, and use of standards and specifications for United States industry.[17] They work closely with similar international organizations such as ISO[18] to harmonize and enhance the global conformity of products to these standards. Both of these organizations are active in the identification and to speed up the development of standards related to growth of additive manufacturing as in the America Makes ANSI Additive Manufacturing Standardization Collaborative[19] and the ASTM Committee F42 on Additive Manufacturing Technologies for the development of AM standards.[20]

As applied to metals, common standard test specimens include *tensile bars*, *Charpy V-notch* tests, and creep specimens. Many a material scientist and materials engineer have spent their technical lives studying the mechanical properties of

---

[15]As specified in Standards, (e.g., ASTM E 1447 or 1409).

[16]The ASTM Web site provides more detail, http://www.astm.org/ABOUT/overview.html,(accessed November 28, 2016).

[17]The ANSI Web site provides more detail, http://www.ansi.org/, (accessed March 26, 2015).

[18]ISO International Standards Organization Web site, http://www.iso.org/iso/home.html, (accessed March 26, 2015).

[19]America Makes ANSI Additive Manufacturing Standardization Collaborative Web site, https://www.ansi.org/standards_activities/standards_boards_panels/amsc/Default.aspx?menuid=3, (accessed November 28, 2016).

[20]ASTM Committee F42 on Additive Manufacturing Technologies, https://www.astm.org/COMMITTEE/F42.htm, (accessed November 28, 2016).

metals. These test methods are actively being researched, modified and applied to AM materials. These active areas of development are leading to identifying the need for new standards to speed the characterization and certification and adoption of AM processes and materials. An introductory description of ASTM standards can be found in student material provided on their Web site.[21]

Other destructive tests that may be applied to AM material could include something as simple as the standard guided bend tests as applied to 3D weld deposits. Guided bend testing is used in weld processing or metal to reveal internal flaws in weld deposits and HAZ regions. Flaws such as voids, cold laps, lack of fusion, poor fusion, porosity, or other microstructural defects can be revealed in surface or near surface regions. Flat plate samples taken from bulk AM coupons may be surface machined and bend tested representative of orientations related to the X, Y, Z orientation of the part within the AM build chamber.

Nondestructive testing of machined test coupons such as those described above may lend themselves more readily to inspection by UT, RT, and PT. While not testing an actual part, these samples can infer a represented characterization of bulk AM material, parameters, and procedures as qualified for an actual build procedure. As you can see from the discussion above, the inspection of AM parts remains a challenge as the procedures and technology are still being developed.

11.5.3 Form, Fit, Function, and Proof Testing

Form and fit of prototype designs was the original function of 3D printed parts. The ability to visualize and handle a 3D part, rather than crowd around a set of blueprints or a computer generated model, has advantages, especially for customers and stake holders not intimately associated with the technical design phase. The same goes for fitting a 3D model into a prototype system. While 3D CAD models can be assembled into virtual systems to simulate part and system function, there is great value in someone actually bolting up a part and confirming they can swing a wrench on it. The prototype does not need to be metal to add value to the form and fit stage of prototype development.

Composite materials are blurring the boundaries of plastic and metal used in prototypes for fit testing and functional testing. Advances in materials offering high temperature and strength, such as metal filled plastic composites, have been applied to offer an increasingly wide range of functional testing. The benefit of these hybrid materials is primarily their lower cost. Metal prototypes go one better by potentially offering the testing of the full functional range of part performance albeit at a greater cost. Metal prototype parts may also be used in short run production such as in molds, dies, tooling, and inserts used to produce other prototypical parts by

---

[21]ASTM introduction to standards, http://www.astm.org/studentmember/Standards_101.html, (accessed March 26, 2015).

conventional means such as injection molding or casting without incurring the high cost of large production run tooling.

Full functional testing of rapidly produced prototypes, saving both time and money, has been the primary function of AM technology. Although many applications are moving into full production prototyping will still demonstrate the technology. Large corporations with highly skilled technical teams are routinely using this technology for larger and larger parts in an increasingly wider range of materials. The boundaries of what the technology can do are being pushed at a very rapid rate. Smaller businesses are finding they need to rely on service providers to tap into this market to stay competitive. Applications are emerging where full production of small lot size components is possible, but for now, keeping up with the competitor's rush to market by using AM is sufficient motivation for small businesses to explore the use of AM. Due to cost and limited availability of engineering skills, AM machines and specialized facilities, small and medium businesses are turning to established service providers to meet their needs. In some cases, large corporations are acquiring or merging with these service providers to establish in-house skills or to deny access to these services by their competition.

Proof testing is a way to subject a prototype part, a process qualification lot or a random sample taken from a production lot, to a mockup testing of the part within an in-service environment. As an example, this is the testing stage where you fire up the rocket engine nozzle, take the car to the race track or pressurize the storage vessel to see how well it works. Often the component is subjected to an extended range of standard operating conditions to prove functional performance beyond that typically seen in service. Elaborate machines to perform these tests have been used for years to evaluate test parts destined for manufacture. We have all seen the so-called shake, rattle, and roll systems where you bolt up the car and subject it to high cycle fatigue for tens of thousands of loading cycles and we all know the fate of the crash test dummies. It probably would not be too long before these machines and dummies will feature 3D printed parts.

11.6   Standards and Certification

Prototype components used for in-house testing and evaluation are often built with a lesser degree of formality and quality documentation that is needed for certification of production parts used in critical. Risks associated with the use of prototype AM hardware built for in-house use can be better managed and controlled than those sent to a third party or customer for testing or use. Prototypes are often fabricated without formal operator training, specifications, standards or certified materials. Costly inspection techniques may not be used at this stage of development.

However, production items sold to the public or introduced into the commercial market need the additional rigor and confidence required by both the producer and consumer. Critical components, or systems of components, such as flight critical hardware used for aerospace, automotive components, energy or medical applications all need to adhere common standards, and meet regulations or codes set by governing bodies. Organizations such as the American National Standards Institute (ANSI) promote, facilitate and approve the development of standards, while organizations such as ASTM International (ASTM), International Organization for Standardization (ISO), and the American Welding Society (AWS) and the Society of Automotive Engineers (SAE) assemble committees of experts to lead the development of voluntary consensus-based standards. Governing bodies that license and certify persons, organizations, products, and services include Federal Aviation Administration (FAA), the Food and Drug Administration (FDA), and the Occupational Health and Safety Organization (OSHA) to name a few. The reader is encouraged to search the Web links of these organizations to learn the latest AM related activities of these organizations.

Regulations, rules, standards, and recommended practices have evolved over decades to apply controls to the production of hardware, feedstock materials, and the qualification of practices and procedures for the benefit of industry, individuals, and the public at large. Certification of metal components used in critical applications and systems is a long and evolved process, with cradle to grave controls and documentation, for materials, qualification of fabrication processes, procedures and operators, and archive of inspection and quality records. As an example, the design of a new turbine blade may cost on the order of $10 M, and certification of the new blade may cost $50 M while the certification of a new system such as an air craft engine could cost $1B. Figure 11.12 shows a "System V" development cycle for the verification and validation of components and systems. This type of development cycle can be modified for a wide range of components and systems with the ultimate goal that maps the process of how *verification* is used to assure part meets the specified design requirements and how *validation* is used to assure the part functions as intended.

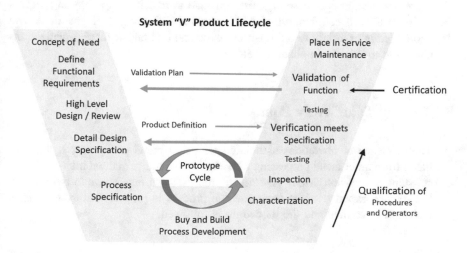

Fig. 11.12   System V Product Life Cycle

All of these organizations are actively involved in developing and modifying certification procedures and processes to accommodate the use of AM fabricated parts for a wide range of new materials and processes. Solving the AM metal part certification challenges of today is many ways lagging behind the pace of technology development, demonstration, and the increasing rate of market acceptance. The economics and cost benefits of moving to the production of certified AM parts are being pulled by demand. There remains, however, a large distance between the production of test coupons, functional test hardware, and fully certified parts. An increasing number of AM parts for use in the medical and aerospace field are being certified for use on a case by case basis. This one-by-one process is costly and time consuming but provides increased justification to identify new industry standards needed to apply AM fabrication to a wider range of components.

For those not familiar with production requirements, here are a few. Some of the formality of operations needed for the move from prototyping to production include equipment calibration, operator certification, development of standards, ISO 9000 quality control, documentation, records keeping, and formal specifications, working to standards and qualified materials operators and suppliers. For those not aware of these requirements, they can easily exceed the cost of design, prototyping, and process development. The costs of launching a product into production can be huge and if the production volumes are low the price per unit must be very high. The certification process may also require additional component builds, and verifying that parts meet all requirements, regulations, and specifications, such as those required for testing and process verification followed by functional validation to assure customer requirements are met. Certification for AM may require automated systems specifically designed and built to automatically characterize AM materials to populate material property databases.

The reader can use the links provided in this chapter and at the end of the book for more detail regarding these organizations and to remain up to date on progress being made toward certification of AM technology. Better yet, if you think you have knowledge or insight to offer, get involved and join one of these organizations to contribute to this world changing effort.

11.7   Key Take Away Points

- The sequence of operations associated with AM metal processing is differentiated from other metal processing operations that require manual intervention or constant supervision. Many AM metal processes can take hours to build a part but require expert oversite to assure the process is properly setup, the safety envelope is verified and the procedure is followed.

- Post-processing operations, inspection, and quality assurance rely primarily on the modification of existing inspection methods to assure a repeatable process, allowable levels of flaws and a component that meets quality standards. The size, distribution, nature, and origin of AM metal defects in some cases challenge existing NDE methods.
- Standards are being developed to assist in the certification of critical components and are needed to fully realize the adoption of AM metal processing in wide-scale industrial production.

Chapter 12
Trends in AM, Government, Industry, Research, Business

Abstract In this chapter, we identify current trends in government sponsored programs, universities, industry, and private enterprise, technology development, and adoption. Taking a higher level view we predict how the technology will connect these sectors and where we will see the greatest impact in the next 5 years. 3D printing and AM technology have been in development for at least the past two decades with the hard work of science and engineering being done at universities, national labs, and within corporate research labs. Advanced manufacturing has seen large investments on the order of billions of dollars driving a higher level of activity and high profile of media attention. The global impact of additive manufacturing is gaining momentum with high levels of funding seen at the government, university, and major corporation levels. Emerging economies are seeing the advantages of developing AM capabilities without the burden of historical infrastructures and as a potential means to leap frog the development of costly infrastructure or creating an advanced manufacturing infrastructure well suited to regional needs. Business, commerce, intellectual property, global, and social issues are discussed and reference leading national and global intelligence reports. The author concludes with a summary of the trends and destinations of highest impact for AM metal technology.

There has been a lot of media attention recently with the infusion of government funding, advances in industry adoption and investment, and the introduction of personal 3D printers all which is raising awareness and expectations. Realists are trying to cut through the exaggerated expectations as shown in the Gartner Hype Cycle (Fig. 12.1). More detailed analysis is available in the report "Hype Cycle for 3D Printing," which is available on Gartner's Web site.[1] It is important to note while widespread adoption and application of AM to these marketing sectors maybe years away, a number of products have made it all the way through the adoption cycle and reached full levels of industrial production.

We agree with the need to keep things in perspective and offer our view of where the technology vectors are pointing. Without a doubt, 3D printing, direct digital,

---

[1]"Hype Cycle for 3D Printing, 2016," which is available on Gartner's Web site at https://www.gartner.com/doc/3383717/hype-cycle-d-printing (accessed January 29, 2016).

© Springer International Publishing AG 2017

J.O. Milewski, *Additive Manufacturing of Metals*, Springer Series in Materials Science 258, DOI 10.1007/978-3-319-58205-4_12

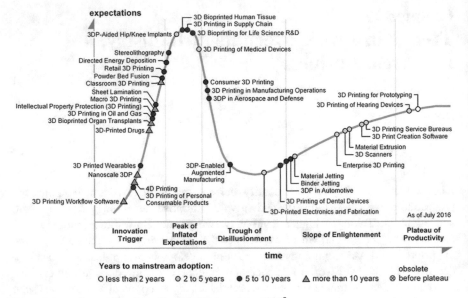

Fig. 12.1 Gartner Hype Cycle for 3D Printing, July 2016[2]

and additive manufacturing of prototypes has crossed the chasm of adoption, the only question is to what extent will mainstream production follow and when.

12.1   Government and Community

Government sponsored programs are being used to stimulate, transition and build economies, and to incubate precompetitive research to foster and assist consortiums to provide vision, mission, and momentum in the national interest. These programs can help consolidate corporate and academic research resources, not otherwise within the reach of individual companies or universities. The recent federal support to AM and advanced manufacturing is being directed to rebuild and revitalize manufacturing bases, stimulate the economy, restore, and stimulate manufacturing job growth.

America Makes (formerly NAMII) in the US,[3] AMAZE[4] in the EU and ARC[5] (Australian Research Council), are three examples. China, India, and the Pacific Rim are following the lead with their host of programs. Substantial funding is being

---

[2]Courtesy of Gartner, reproduced with permission.

[3]America Makes Web site, https://americamakes.us/, (accessed March 26, 2015).

[4]AMAZE project Web site, http://www.amaze-project.eu/, (accessed March 26, 2015).

[5]ARC, Australian Research Council Web site, http://www.arc.gov.au/, (accessed March 26, 2015).

Fig. 12.2 Technology
Readiness Levels[6]

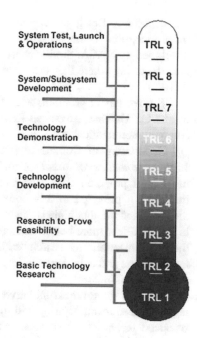

made by the US government over a wide range of agencies and programs at levels
of $10's of million dollars a year, particularly, since 2012. Funding have been seen
outside the US and is gaining momentum. This funding has been creating a lot of
press and, in some cases, wild enthusiasm in the business and finance sectors.
Programs such as these allow graded membership costs and benefits supporting
development of technology and encouraging participation of both large and small
businesses and universities. Government support to students and industrial intern-
ships, supported by in-kind funding toward membership by companies, are making
possible a wide range of participation.

As an example in the US, the America Makes program is funding a wide range
of technology development activities that focus on overcoming the engineering
challenges to advance 3D printing and AM technologies. A few examples include
the development of materials for AM, scale up of AM systems in size and speed,
the integration of AM hardware into hybrid machines, the integration of in-process
real- time quality assurance systems and the creation of new computational models
and AM process databases. The benefits include technology spin off, the forging of
partnerships large and small, creating a larger experienced workforce, promotion of
the technology, and creating a pipeline of STEM (science technology engineering
and math) talent within the workforce.

---

[6]Courtesy of NASA.

Technology Readiness Levels are a way to rate the readiness of a technology from inception to final use and adoption. Figure 12.2 shows one such chart developed by NASA. Similar charts exist for other technology sectors as well as for manufacturing where they are often referred to as Manufacturing Readiness Levels (MRLs).

US government programs and studies associated with AM are being sponsored by the US Defense Advanced Research Program Association (DARPA), the U.S. National Aeronautical and Space Administration (NASA), NIST, US Navy, and the US Department of Commerce (DOC). One example, NIST Special Publication 1163, *Economics of the U.S. Additive Manufacturing Industry,*[7] highlights a number of pros and cons associated with 3D printing summarizing some of the information provided in this book. The good news is this information is being disseminated at the highest levels of government in an accurate manner. I'll highlight a few more below as examples, but the best bet is to follow the links to these organizations and search "additive manufacturing" or "3D Printing" to get the latest reports and calls for technical proposals and collaborations for the most up to date information.

Other US government initiatives and programs focused on advanced materials and advanced manufacturing will integrate into AM technology and provide a more balanced technology landscape. Although 3D printing has been getting a lot of media attention, luring investors with its siren's song, it is pulling the discussion of the benefits and potential of AM along with it, attracting young people and a wider cross section of industrial interest. Advanced materials and advanced manufacturing and digital factories are productive areas of research and development. Although garnering less attention, these fields are interconnected and are being pulled along with the current interest in 3DP. Silicon Valley is realizing this as the infusion of information and information access into products such as cars and personal items is accelerating. Additional government programs with ties to AM include the Lightweight Metals Government initiative,[8] National Center for Defense Manufacturing and Machining,[9] the Materials Genome Initiative,[10] and MGI Strategic Plan.[11] Other suggested reading includes the International Trade

[7]NIST Special Publication 1163, *Economics of the U.S. Additive Manufacturing Industry,* Douglas S. Thomas, Applied Economics Office Engineering Laboratory, August 2013. http://nvlpubs.nist.gov/nistpubs/SpecialPublications/NIST.SP.1163.pdf, (accessed May 14, 2016).

[8]Lightweight Metals Government initiative, http://lift.technology, (accessed March 27, 2015).

[9]National Center for Defense Manufacturing and Machining, http://ncdmm.org/, (accessed March 27, 2015).

[10]NIST Materials Genome Initiative web site, http://www.nist.gov/mgi/, (accessed March 27, 2015).

[11]National Materials Genome Initiative strategic Plan, final version, June 2014, http://www.nist.gov/mgi/upload/MGI-StrategicPlan-2014.pdf, (accessed March 27, 2015).

Fig. 12.3 Future Trends
Intersecting Additive
Manufacturing

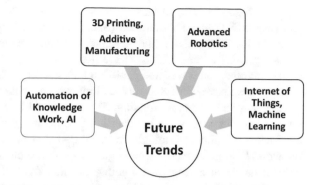

Association, SMI (Sustainable Manufacturing Initiative),[12] which provides an online Sustainable Manufacturing Metrics Toolkit[13] that provides metrics to measure and improve the environmental performance of manufacturing facilities and products.

A recent US DOC Department of Commerce study[14] lists four technology trends: Automation of Knowledge Work, Internet of Things, Advanced Robotics, and 3D Printing and references a recent McKinsey Global Institute study[15] (Fig. 12.3). This study refers to the "democratization of manufacturing," where individuals with 3D printing machines and designs produce their own goods at home.

The Science and Technology Policy Institute provides is a good source of white papers describing the case for the US government's role for investment in advanced manufacturing, such as *Advanced Manufacturing Questions Prepared for the Advanced Manufacturing Workshop of the President's Council of Advisors on Science and Technology's Study on Creating New Industries through Science, Technology, and Innovation.*[16]

---

[12]International Trade Association Sustainable Manufacturing web site, http://trade.gov/competitiveness/sustainablemanufacturing/, (accessed March 27, 2015).

[13]Sustainable Manufacturing Metrics Toolkit, http://trade.gov/competitiveness/sustainablemanufacturing/metrics.asp, (accessed March 27, 2015).

[14]US DOC Department of Commerce Study, 4 Hot Technologies Will Drive Manufacturing's Future, Friday, October 24, 2014, David C. Chavern, https://www.uschamber.com/blog/4-hot-technologies-will-drive-manufacturing-s-future, (accessed March 27, 2015).

[15]McKinsey Global Institute, Disruptive technologies: Advances that will transform life, business and the global economy, May 2013, by James Manyika, Michael Chui, Jacques Bughin, Richard Dobbs, Peter Bisson, and Alex Marrs, http://www.mckinsey.com/insights/business_technology/disruptive_technologies, (accessed March 27, 2015).

[16]White Papers on Advanced Manufacturing Questions Prepared for the Advanced Manufacturing Workshop of the President's Council of Advisors on Science and Technology's Study on Creating New Industries through Science, Technology and Innovation, April 5, 2010. Science and Technology Policy Institute, Washington, DC 20006, http://www.whitehouse.gov/sites/default/files/microsites/ostp/advanced-manuf-papers.pdf, (accessed March 27, 2015).

In the past, US national labs, such as Los Alamos National Laboratory (LANL) and Sandia National Laboratories (SNL) took a leading role in developing AM technologies (Lewis and Schlienger 2000), but now other US National Laboratories including ORNL[17] and LLNL[18] have taken the lead.

All this attention is well and good, but let's not forget the recent history of the past century as we have witnessed surges of interest in technology due to the infusion of government and corporate funding. The boom and bust cycles of artificial intelligence (AI), robotics, or the big splash introductions of microwave cooking, superconductors, or other technologies were all touted as having the potential to "change the world." Some of these technologies may indeed change the world, but in some cases it has been a long time coming, in other cases it has done so, but quietly. Regardless of all the media and investment expectations, 3D printing and additive manufacturing will take its place in the world while advances are being made in other rapidly advancing manufacturing technology sectors as well.

The AMAZE project (Additive Manufacturing Aiming toward Zero Waste & Efficient Production of High-Tech Metal Products) in the EU (European Union) has the goal to rapidly produce large defect-free 3D printed metallic components and to achieve 50% cost reduction for finished parts. €20 million will be invested in this project. Pilot-scale industrial AM factories are being set up in France, Germany, Italy, Norway, and the UK to develop the industrial supply chain. The 28 partners include Airbus, Astrium, Renishaw, Volvo Technology, Norsk Titanium, Cranfield University, EADS, the University of Birmingham and the Culham Centre for Fusion Energy.

The European Space Agency (ESA) features other such projects.[19] A good EU roadmap, identifying projects, players and needs, can be found at this link,[20] with the EU Diginova project focusing on innovation for digital fabrication[21] and the EU Merlin project focusing on the development of Aero Engine Component Manufacture using Laser Additive Manufacturing.[22]

Australian AM interest may be found in the Australian Additive Manufacturing Technology and Cooperative Research Roadmap.[23]

---

[17]Oak Ridge National Laboratory, US department of Energy, Innovations in Additive Manufacturing, web page, http://web.ornl.gov/sci/manufacturing/research/additive/, (accessed April 6, 2015).

[18]Lawrence Livermore National Laboratory Additive Manufacturing web page, https://manufacturing.llnl.gov/additive-manufacturing, (accessed April 8, 2015).

[19]European Space agency press release, October 3, 2013), http://www.esa.int/For_Media/Press_Releases/Call_for_Media_Taking_3D_printing_into_the_metal_age, (accessed March 26, 2015).

[20]Additive Manufacturing in FP7 and Horizon 2020, report from the EC Workshop on Additive Manufacturing held on 18 June 2014, http://www.econolyst.co.uk/resources/documents/files/EC_AM_Workshop_Report.pdf, (accessed March 26, 2015).

[21]EU Diginova project, www.diginove-eu.org, (accessed March 26, 2015).

[22]MERLIN project Web site, www.merlin-project.eu, (accessed March 26, 2015).

[23]Australian Additive Manufacturing Technology and Cooperative Research Roadmap, Martin, J., February 28, 2013. http://amcrc.com.au/aamtsummary, (accessed March 27, 2015).

China[24] and India are following suit with programs funded to establish AM technologies and foster the growth in fields such as aerospace, transportation, and the medical field. A recent article in the Economist[25] cites the growth of AM to include companies producing AM parts in metals such as Beijing Longyuan Automated Fabrication Systems (known as AFS). A recent article in Engineering.com cites the growth of additive manufacturing in China, asking if the US is losing its edge.[26] Additive manufacturing is taking hold in India with the Additive Manufacturing Society of India[27] promoting 3D printing and additive manufacturing technologies.

More recently, the governments of Japan and other Pacific Rim countries are providing greater investment and seeing greater corporate research and adoption through global business partners. Initiating the investment and establishment of AM manufacturing centers across the globe can only help all participants as the work force, regulatory environment, market places, and economic needs differ significantly enough that each participant may lead in some areas of technology and business application while following and learning in others.

Perhaps in the future, these government and corporate sponsored programs will provide free engineering quality cloud and software resources to serious students and apprentices. In some cases, support may take the form of live-in, work-in environments, where room, board, and hardware resources are made freely available to support serious self-learners and makers.

The current trends in government support are favorable but what could happen to slow or prevent this growth? Certainly the US, as a leader, has had economic advantages due to recovery from the recession and strong corporate growth. To its credit, the Obama administration prioritized manufacturing program funding, attracting corporate participation. A change in administration and a reversal of these polices could slow growth, but a foothold in AM has been established and will certainly not experience a reversal.

The establishment of a national database of AM designs and processes is needed to allow uploading, sharing and downloading. Some designs and product functions may be most appropriately distributed in a public and open source forums, such as government sponsored research or designs resulting from government funded programs. Government incentives to create and submit designs or process data to realize these designs need to be developed. These government incentives may take

---

[24]Rapid Ready for Design Engineering article, "China to Invest in Additive Manufacturing", Posted by John Newman, December 18, 2012. http://www.rapidreadytech.com/2012/12/china-to-invest-in-additive-manufacturing/, (accessed March 26, 2015).

[25]Economist article, "A New Brick in the Great Wall", April 27, 2013, http://www.economist.com/news/science-and-technology/21576626-additive-manufacturing-growing-apace-china-new-brick-great-wall, (accessed March 26, 2015).

[26]http://www.engineering.com/3DPrinting/3DPrintingArticles/ArticleID/5716/US-Losing-its-Edge-in-Additive-Manufacturing.aspx, (accessed March 26, 2015).

[27]Additive Manufacturing Society of India Web site, http://amsi.org.in/, (accessed March 26, 2015).

the form of free access to government sponsored tools, such as free credits for use of software or access to hardware such as scanners or printers or government data bases. Tax credits may also be granted to encourage contribution. Contribution may appropriately be regulated to help control distribution or quality. As an example, commercial contributors and consortium members engaged in precompetitive research may be granted preferential access prior to full open release. Limitations or conditions may be made to the release of beta-grade content awaiting validation or substantiation.

State governments and universities have joined with the support of federal funding to tie federal, corporate and university activities into communities and serve as a conduit to local educational and business resources. One such example is the Center for Innovation in Additive Manufacturing at Youngstown State University[28] offering education up to the Ph.D. level, and workforce development opportunities.

At the local level, city and community support has grown to include educational opportunities and activities at community colleges, trade schools, and other organizations. Perhaps forward looking federal, state, or local governments can offer tax breaks, subsidies, or other compensation to companies offering local or remote print services featuring educational discounts.

Maker clubs, fab, and tech spaces and community resources, such as community colleges and technical institute programs, can serve as learning and innovation centers. Pooling of resources will allow participation, even at modest levels. Virtual maker spaces may exist to leverage widely distributed networks of specialized skills. It is not beyond possibility that inspiration/idea, CAD model building, CAM build path planning, FEA analysis, specialty material production, even monitoring, and control could exist far from the actual machine resource used to build the part. The Fayetteville Free Library's Fabulous Laboratory[29] is one such example, where free and open access to 3D technology is used to foster creativity, education, and to provide resources for people to make things. Another example of a digital education space is the Brighton Maker Lab.[30] Open sourcing, sharing, pooling or in-kind services, skills, and IP could further connect maker group members to build environments at a reasonable cost to those who would otherwise be unable to maintain all those resources in one space. Wide geographic dispersion and crowd source cooperation will be made possible by open resources, membership, and sharing.

Commercial Web-based fabrication clubs are already realizing and extending these forums for making, learning and invention. Examples include Fab Labs[31] and

---

[28]Center for Innovation in Additive Manufacturing at Youngstown State University, http://newsroom.ysu.edu/ysu, (accessed March 27, 2015).

[29]Fayetteville Free Library's Fabulous Laboratory Web site, http://www.fflib.org/make/fab-lab, (accessed March 27, 2015).

[30]Brighton MakerLab web page, http://makerclub.org/makerlab/, (accessed March 27, 2015).

[31]Fab Lab Web page, https://www.fablabs.io/, (accessed March 27, 2015).

the FabFoundation[32] which provide access to modern means for invention. The Maker Movement will continue to draw young people into the industrial arts to experience the satisfaction and rewards of working with your hands. Quality of life vs. quantity of life will reflect itself in a sustainable work existence where the value in one's work is realized more by excellence in creation than by the relentless pursuit of consumption and accumulation. An inquiry into work values is provided in Shop Class as Soulcraft (Crawford 2009). Maker communities are getting corporate support as well with leaders such as Autodesk[33] and Intel[34] expanding access to innovative design tools.

Contests and competitions are a great way to motivate teams of cutting edge thinkers and makers. As an example, government sponsors, such as DARPA and NASA, have held large-scale competitions for autonomous vehicles and robotics attracting the best and brightest teams from academia and industry. Large corporate sponsors such as GE, partnering with GrabCAD and NineSigma,[35] are offering global competitions, such as *GE Quest*, to attract and crowd source innovation. The *GE Innovation Manifesto* features the principals of transparency, public evaluation criteria, compensation, IP rights to pool IP, and knowledge into a concept they refer to as the *Global Brain*. Other commercial entities are sponsoring contests with smaller prizes targeting other technical and artistic demographics with interest in the technology and access to open resource tools.

Open source thinking spaces with makers tinkering in garages, shop spaces and club meetings are springing up everywhere. While a maker may not easily build a new electron gun or run zero gravity experiments at home, many of the present and future challenges for AM technology could benefit from innovative out-of-the-box thinking, both by individuals and crowd sources focusing on specific or highly generalized solutions. The capture of the public interest and imagination, empowered by public maker resources, open source information, free software, online learning and information sharing will help propel this and other AM technology advances.

---

[32]Fab Foundation, http://www.fabfoundation.org/, (accessed March 27, 2015).

[33]PSFK Labs, Intel IQ, "Expanding 3D Printing into a Maker's Repertoire", by Nora Woloszczuk, http://www.psfk.com/2014/10/expanding-3d-printing-makers-repetoire.html, (accessed March 27, 2015).

[34]IQ Intel, PSFK Labs, "Digitizing The Analog: The Ever-Expanding Toolkit For 21st-Century Makers", http://iq.intel.com/digitizing-the-analog-the-ever-expanding-toolkit-for-21st-century-makers/, (accessed March 27, 2015).

[35]Nine Sights Community, www.ninesights.com, (accessed March 27, 2015).

Small businesses and innovators will continue to leverage precompetitive research, open-source publication and collaborative research by consortium membership, and to launch products-and services-based upon newly developed AM technologies. Small business research grants and small business access to membership into diverse government led programs will help to prevent monopolization of technology segments by industrial giants.

Education pipelines are being formed to interest and train a new generation of makers, techies, and budding engineers. Starting in K-12, learning programs may be offered across schools and community groups to leverage the fun of creation into social and team-thinking settings. A trend to build objects (think LEGO League) will undoubtedly be applied to modeling software, 3D gamify design software, as well as making, printing, and mobile applications. Communities will offer free access to open source software and hardware resources as an educational benefit to registered students.

Group learning and virtual craft guilds will spawn from TechShops and maker spaces and cloud computing services will allow pay by usage, versus flat rate, or membership rates. Membership fees for eCo-ops or eCraftGuild may become dependent on your contribution to the body of information, resource pool, or cloud content. Certified craftsmanship status may be attained as your designs or methods are downloaded and used. Micropayment and micro financing may be offered directly or indirectly to content providers.

12.2   University and Corporate Research

Universities across the globe, often supported by government or corporate funding, are continuing to perform precompetitive and sponsored research needed to identify and address challenges from all aspects of AM including materials, software, modeling, sensors and hardware integration and process development. Open source journals and leading scientific journals are featuring an impressive array of new AM technical work. Google Scholar is a good place to see who-is-who and what they are doing in groundbreaking research and innovation. Titles, authors, abstracts, patents, and references help identify the most active areas of research and who is doing it. Setting up a Google Scholar alert is a great way to keep abreast of scientific and technical advances. Open-source technical papers and Ph.D. dissertations are sources to digging deeper into the challenges faced, the questions being asked, those answered, and proposed solutions. An excellent open source of technical papers is the International Solid Freeform Fabrication[36] Webpage. This

---

[36]International Solid Freeform Fabrication Web page, University of Texas at Austin, http://utwired.engr.utexas.edu/lff/symposium/proceedingsArchive/, (accessed March 27, 2015).

source provides a historical time line in the evolution of AM technology and identifies the technical pioneers and current leaders of AM processing technology.

- A wide range of engineering schools offers scientific and technical degrees up to Ph.D. level. It is unfair to create a list, as many excellent schools are offering programs and the list is changing and evolving as rapidly as the funding. A few university sources I found helpful in writing this book include the University of Texas at Austin, Pennsylvania State University CIMP-3D, University of California—Irvine, the University of Louisville, the University of Texas—El Paso, and Pennsylvania State University, Applied Research Laboratory, University of Nottingham, and those identified in the footnote links.
- A listing of government, university and corporate participants may also be found on the America Makes membership list.[37]

Research on first principles based physics models will continue to be developed as better material property data, up to melting and vaporization, improve accuracy, and software and computing hardware advances speed computation and improve resolution (Metals Process Simulation, ASM Handbook 2009) (Furrer and Semiatin 2009).

Materials databases and microstructural models will provide a better understanding of the process-driven metallurgical response while large databases and data-driven solutions will provide information associated with the complex cause and effect relationship between input and output parameters helping to optimize parameter selection. Inverse, reduced order, and data-driven models, combined with real-time monitoring and control, will further optimize AM speed and quality. Multi-scale models will link molecular, metallurgical, and bulk scale responses with process conditions and ultimately to part and system performance.

Computer simulation technology being developed at the Universities is beginning to move outside academia and into the private sector. In one example, 3DSIM (Fig. 12.4) is offering software and technology services to predict residual stress, distortion, and to help design AM process and support structures utilizing efficient algorithms and advanced computational methods.[38]

---

[37]America Makes membership list, https://americamakes.us/membership/membership-listing, (accessed March 27, 2015).

[38]3DSIM website, http://3dsim.com/, (accessed August 13, 2015).

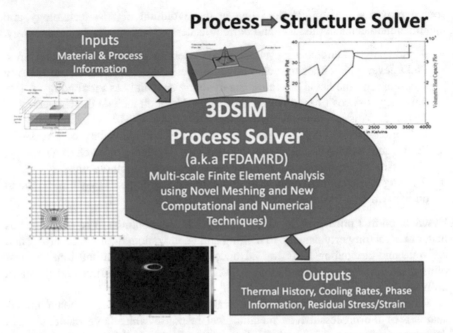

Fig. 12.4   3DSIM Process Solver[39]

Other government sponsored initiatives such as the Material Genome Initiative,[40] will have links to AM. Corporate research is most often focused on applications, materials, designs, processes and the value stream most relevant to their line of work. Current engineering applications in corporate research labs are often directed and constrained by corporate focus, corporate culture, and the evolutionary needs of the business. A number of large global corporations are taking bold moves to establish leadership positions, sharing their vision, and demonstrating their commitment with huge investments, while accepting the risks of early adoption. Others are following and joining in while evaluating the various technology options, positioning themselves in a safer position on the adoption curve, while remaining visible as players in the "AM game."

As mentioned above, large corporations are gaining further attention, identifying and attracting new talent by sponsoring contests and competitions. In one such example, Layerwise won the GE sponsored 3D Printing Production Quest innovation challenge[41] to produce a complex medical imaging device using a refractory metal.

---

[39]Courtesy of 3DSIM, reproduced with permission.

[40]Materials Genome Initiative, final draft June 2014, https://www.whitehouse.gov/sites/default/files/microsites/ostp/materials_genome_initiative-final.pdf, (accessed March 27, 2015).

[41]Medical Design Technology article, "LayerWise Wins GE's Global 3D Printing Production Quest", 06/04/2014, by LayerWise, http://www.mdtmag.com/news/2014/06/layerwise-wins-ge%E2%80%99s-global-3d-printing-production-quest, (accessed March 27, 2015).

Corporate participation in consortiums with nonprofit technical organizations, universities, businesses, and government agencies offer another way to leverage member and government funds and contributions to advance AM technology. The Edison Welding Institute (EWI), in Columbus, Ohio is one such nonprofit and has formed the Additive Manufacturing Consortium[42] (AMC) with a mission to bring together AM users to foster technical interchange, execute group sponsored projects, collaborate on government funded opportunities and roadmaps.

In some cases, established companies are finding themselves behind the curve in thinking how these new technologies will come into play and ultimately how they will do business in an evolving market place. Investment details and legal aspects related to patents, copyrights, and trade secret information will all come into play. Agile newcomers will jump into the game and leap-frog ahead, unburdened by conventional factory infrastructure and historical capital investments. Some will die along the way. Others may stake a claim to the new technology with a modest investment, while awaiting the shake out of participants and the demonstration of clear value-added pathways. Some companies are undoubtedly establishing their own AM printing capability as a way to protect intellectual property by keeping prototype fabrication in house.

Professional Societies are continuing to sponsor AM-related conferences and committees to bring together users and developers of AM technology, and to create committees to propose and harmonize domestic and international standards, practices and terminology. Rather than providing outdated information, the Professional Societies and Organizational Links section of the book has a listing the reader can go to and search for the latest AM conference, workshop or committee activity. In some cases, adoption and recognition of AM technology has been slow and deliberate. As an example, despite a 20 year history of AM development, the Metal Powder Industry Federation[43] has waited until 2014 to sponsor its first AM Uses of Powder Metals conference, displaying the caution with which entire industries move toward the adoption of fundamentally new technology.

Professional society interest and support is an important component to sponsor forums for recognition, precompetitive sharing of ideas and applications. As an example, this Minerals Metals and Materials Society (TMS) forum[44] addressed key issues related to powder processing, AM manufacturing using high-performance materials and new AM processing methods.

---

[42]Edison Welding Institute Additive Manufacturing Consortium, Web link, http://ewi.org/additive-manufacturing-consortium/, (accessed March 27, 2015).

[43]Metal Powder Industries Federation Web site, http://www.mpif.org/, (accessed March 27, 2015).

[44]The Minerals Metals and Materials Society, 2012 Near Net Shape Manufacturing Workshop, April 11–13, 2012, Moline, Illinois, USA, TMS, http://www.tms.org/meetings/2012/NearNetShapeManufacturing/home.aspx, (accessed March 26, 2015).

In another example, a team of SME advisors, in cooperation with the Milwaukee School of Engineering and America Makes, has strategically defined the additive manufacturing body of knowledge to serve as the basis for an Additive Manufacturing Certificate Program.[45]

12.3   Industrial Applications

Industrial trends in AM primarily focus on adoption, establishment, and optimization of the various direct or support technologies employed by suppliers, users, and those supporting the technology infrastructure. As stated above, these trends often are linked and follow government funding, corporate R&D, or university research and provide leverage or pull into the marketplace.

Industrial users and AM machine builders acknowledge the relentless need to offer improvements to process speed, accuracy, surface finish, material properties, quality, and repeatability. The development and improvement of new laser technology, electron guns, multiple beams, faster powder spreading, and fully integrated modeling and optimization and print software is seen each year at product shows. Innovative solutions are rapidly evolving to reduce the cost and improve the quality of powders, wire materials, and to optimize the value chain.

Databases relying on information gathered directly from production machines, in-process monitoring, embedded sensors, wireless connections, internet, and cloud resources will generate big data, bigger databases and ultimately data-driven solutions. Algorithms to analyze this data will provide insight above and beyond the points of data origin by fusing and mapping this distributed source data into a better understanding of all processes. This will of course depend on the participation of quality data providers while addressing IP and proprietary data concerns. Security concerns may be overcome by further developing methods such as secure anonymous-contribution membership status for donations to the data bank in return for membership-level database access.

Highly hybrid digital work cell systems will integrate AM with advanced CNC and NDT inspection systems to provide work cells with unprecedented capacity and flexibility. In addition to laser powder fusing, these system may offer laser drilling, laser cutting, laser glazing, laser assisted machining, engraving, scanning, and measurement. Fully integrated mobile transport container sized work cells will be delivered by land, sea or air, to be controlled by remote wireless link to the cloud, requiring minimal on-location support.

Data matrix codes and IDs fabricated onto surfaces by AM will allow the reading of codes in remote or harsh environments where current marking or reading technology will not function. These Braille-like codes, built along with the part, will

---

[45]SME AM Certificate course link, http://www.sme.org/rtam-certificate-program/#details, (accessed April 8, 2015).

enable linking to component designs, specifications, manufacturer information, and in-service records. They will allow the addition of inspection reports or in the case of component failure, a direct request for the rapid delivery of a new part to a specified maintenance pickup point.

Large motion system accuracy decreases and cost often increases with the size of the build volume (cubic scaling). High precision drive mechanisms scale in cost much more rapidly than the controllers and motors needed to move them. This trend will slow as new technology is developed to serve the demands of industry. Accuracy requirements for very large build volume motion will seek advanced solutions. Dynamic accuracy mapping will evolve with software compensation controlling position in real time for CNC systems.

Expandable or configurable build volumes for powder based systems will be developed to resize the machine for parts of specific geometry or high aspect ratio components. This will reduce the total volume of powder required and avoid the resource constraints of cubic scaling.

Industrial standards and procedure specifications will emerge, linked to process qualification records needed to produce certified product definitions to allow the *next-shore* production of functionally identical articles on a wide range of geographically dispersed fabrication platforms.

Cyber security of files and transfer of information will be protected by certification algorithms in software or at the chip level using encrypted verification codes or data encryption schemes to assure digital product definition or process data was not corrupted, hacked, or violated during generation, storage, transport, or use. Standards and verification methods will be in constant evolution as the cat and mouse game of IT security will never end.

The capacity to handle *Big Data* (V.C. Mayer-Schönberger 2013) will extend to the individual component performance level and not just the industrial systems level. Today, thousands of distributed sensor networks monitor everything from pipe lines to jet engines. This will be extended by the Internet of Things (IoT) to all components. AM will play a role in how sensors are embedded in metal objects. This relentless level of data taking and refinement will enable the capability and capacity to measure and connect everything. Analysis of Big Data will lead to discoveries and optimization all along the value stream.

Cradle to grave "hardware product DNA" will be mapped and encoded in all virtual product definitions to include information related to material, product definition, process definition, test, performance, and in-service data. In-service data will be used to modify, improve, renew, repair, or replace components.

Evolved hardware product DNA will include variants and evolutions of the base product DNA to include customized designs evolved from in-service data, e.g., including failure modes or wear patterns or to accommodate custom service conditions and performance data. One hypothetical example is an engine exhaust manifold design modified (evolved) to use in one application then modified to include a different material for a corrosive environment (e.g., sea water) or strengthened by a thicker section for a high shock (e.g., off-road) service environment.

Reconstructed product DNA may become a commodity, as historical objects with no surviving design information, or with only a paper design definition, are converted to 3D CAD models for AM use. Intelligent systems capable of reading blueprints and converting them to 3D models will be developed. Patent and copyright laws will need to accommodate this and other reverse engineering considerations. Retired design engineers may be paid to contribute legacy information regarding original design intent before it is lost forever.

Micropayments or information credits may be offered to encourage the reporting of data related to AM production or AM component in-service use. Components needing maintenance, repair, or remanufacture may be predicted or identified. Payout or redemption of data credits may scale with the quality or volume of the information provided.

Highly hybridized designs will be developed, where multiple parts, part functions, or build parameters may be combined into a single optimized design. The actual build cost increment will be small or nonexistent while the performance benefits may be significant. Component designs will focus on function rather than upon manufacturing methods.

Validation of designs, processes and products is needed to ensure safety, quality, and IP protection. Efforts will be ongoing, both within corporations and within consortiums to share data along the value chain. Government sponsored programs will push the continued creation of open-source databases and standards, to encourage small-and mid-sized businesses to adopt new AM technology and to allow wider adoption outside of proprietary corporate R&D efforts.

Retailing of products in the cloud is already being realized by large corporate entities, such as Amazon, Walmart and UPS. As retail sellers and distributors, these companies are demonstrating belief and investment in AM technology all along the supply chain. Web-based product offerings may soon offer virtual 3D printed products alongside conventionally produced and warehoused items potentially at a discount or with user-defined attributes or features. This will begin with plastic and polymer items but evolve to include metals, composites, and ceramics.

The warehouse in the cloud concept will evolve to contain items both within distribution warehouses as well as within storage warehouses. Product and process definitions will be created for machines and systems in service starting with maintenance, repair, and upgrade items. With the ubiquitous expansion of high-speed data access to more of the planet, designs for most anything will be available most anywhere with access to a AM machine. Definition of molds and tooling will be stored in virtual warehouses to serve the tail end of the market for replacement parts serving small lot, short run, or even single part needs. In addition, a warehouse of virtual fixtures, manufacturing aids and tooling will further reduce storage and inventory costs to serve the demand for conventional fabrication methods as applied to rarely requested replacement parts.

Fabrication sequence translators, similar to those in existence for translating 3D model file formats (e.g., STEP, IGES, STL), will be developed as the subsequent generations of AM machines and software tools are retired or made obsolete. Functionally equivalent substitutions will be needed for the redevelopment of fabrication processes previously developed using proprietary materials, processes, machine configurations, algorithms or controls having made direct transportability of process conditions impossible. The function of these translators will mitigate the need to step all the way back to the original product definition to completely redevelop, requalify, or recertify the entire process. These translators will be able to reuse an amount of historical information, beyond the original the product definition, (existing product DNA) to speed the recertification of components built on different or evolved equipment of subsequent models, technology generations, or modifications.

Companies will scrutinize their databases of designs to identify candidates for digital retirement of designs, tooling and fixtures and to reduce warehouse storage requirements, in essence transferring them to digital warehouses.

As stated earlier in the book, model and enterprise based software such as CAD, CAM, CAE, and CNC will continue to need revision and customization to provide greater utility for AM and across the entire life cycle of a product from design to end of service life. Specialty software will provide the specific utility to engineering applications. PLM (Product Lifetime Management) and ERP (Enterprise Resource Planning) software, to include factory simulation and service network scheduling will need revision as historical value and IT chains evolve to include AM.

Development of industry standards will continue to race with technology development to enable the adoption of AM into wider industrial products. Standards such as ASTM F3049 Guide for Characterizing Properties of Metal Powders Used for Additive Manufacturing Properties or ASTM WK43112 Guide for Evaluating Mechanical Properties of Materials made via Additive Manufacturing, will undergo many revisions during the maturation of the technology.

The benefits of creating designs for freeform fabrication and process planning will be expanded by a new generation of engineers. Historical constraints to AM adoption will be relaxed as they will no longer be tied to a culture of conventional fabrication and design thinking. Automated systems for planning and control will optimize the selection of machine parameters (speed, laser power, powder flow, etc.) based upon data-driven algorithms and the growing databases generated in industry. The aesthetics of designs will also be enhanced without a penalty for complexity or the constraints of commercial shapes or conventional processing techniques.

Process optimization for design and machine function will more widely utilize finite element analysis, evolving toward the concept of finite element fabrication, where high-resolution thermal and mechanical models predict conditions such as full fusion, potential defect location, heat buildup, microstructural evolution, and ultimately will predict quality and performance at every location in the part. Tuning and optimization of process parameters will further improve the quality of the process plan and ultimately the part itself. Microstructural evolution models and alloy design models will be employed for localized tailoring of microstructures and properties within the component.

Software to assist in the fundamental part design, such as by topology optimization, will begin with a minimum set of requirements, input to simulation and analysis software, to realize an optimized design. Forgetting how it was done in the past, or at least not leading with historical design concepts, will be the hard to overcome for designers experienced in conventional/historical processing methods.

The new paradigm for solid freeform fabrication will evolve over the next decade as industrial engineering and applications develop and push the technology. Novel applications and needs, identified by independent designers and makers, will help pull the technology into greater adoption. The ubiquitous application of plastic prototyping will help lead the way for metal AM processing. AM/SM hybrid systems will proliferate in the next decade. The unique properties of new metal alloys, designed for AM processing, will enable applications that cannot be served by plastics, ceramics, or composites.

Repair and upgrade of machine tooling for conventional processing to include AM capability will evolve significantly as the design of components and system designs increasingly become model based. Modular AM systems will be integrated with SM to renew, upgrade, and repurpose machine tools to significantly higher functionality beyond original specifications. As an example, the service life of a punch and die set will be increased by resurfacing and by upgrading the process, materials, and wear features during a rework cycle.

A one-size-fits-one level of customization and consumer-matched products will build what you need, when you need it, where you need it, made specifically for you. This is already beginning to happen with medical and dental implants and crowns. What if these were taken a step further and the item was fabricated locally in anticipation of need, as in a hip joint needing replacement? Mobile SM/AM fabrication labs will localize product fabrication in response to demand for *bespoke* products.

Mobile hybrid systems incorporating AM, laser, machining, CNC and in-process inspection and real-time control will allow maintenance and repair in situ for an ever growing number of applications. Small lot manufacturing centers will be driven or air dropped to remote locations and disaster zones to speed recovery of important services and infrastructure. Direct access to the cloud for original design or fabrication specifications combined with hand-held in situ laser scanning, XRF (energy dispersive X-ray fluorescence) or 3D photo reconstruction will allow determination of damage or wear conditions. Establishment of localized 3D repair resources and machine generated repair sequences will enable on-site fabrication and repair minimizing the need for skilled human intervention. As an example today, the repair of a bearing surface on a marine shaft or rebuilding a broken or worn ship propeller are already being performed on site in a less automated manner. AM will take repair complexity to a new level and new locations such as underwater, at sea, in nuclear exclusion zones, or in orbit.

An ever increasing reliance on the life extension of existing industrial tooling such as custom punches, dies, swage hammers, etc. will occur. As the decades progress, there will be a slowly decreasing demand for outdated parts for which this tooling was developed, making the payback for producing new replacement tooling economically unfeasible. This decreased demand will result in the need for lower cost repair, refurbishment and remake of tooling sufficient for small lot production. Very small lot sizes or single article production may justify direct AM fabrication of the part itself.

As specialized hybrid AM/SM machines are relied on to build just-in-time tooling, the economic justification for storing old tooling and dies on the shelf for years and years will decrease. Scanning, reverse engineering, and translation of paper designs into CAD models will allow the creation of virtual designs of historical tooling warehoused in the cloud, ready for production, as driven by demand. Digital warehousing will increase, minimizing the need for large physical warehouse inventories of spare parts, historical tooling, and manufacturing aids.

At the retail level, the concept of hardware store in the cloud may accommodate needs for uncommon or customized hardware. This may begin with customized plumbing as an example, or ultimately extend to customized tools, replacement parts, or consumer products such as a bicycle that fits one user's leg, arm reach, and grip size.

AM fabrication processes and job shops will run unattended 24/7, requiring minimal human intervention and support moving toward the goal of a lights-out factory. Automation of knowledge work will increase, shifting design decisions away from engineers to analytic software tied to growing manufacturing databases and engaged in continuous machine learning and data-driven optimization.

The integration and consolidation of function and design will allow one part to serve the function of many. Complex design will replace complex assemblies that historically relied upon discreet component fabrication, assembly, and fasteners to achieve the overall function. Multifunctional design and hybrid components will realize complex built-in functions with less assembly required to achieve the same function.

Advanced AM process models and simulation will be developed, based upon the combination of inverse, direct, and data-driven models. Errors, omissions, knowledge gaps, noise, disturbances, fragmented, or incomplete data will be compensated by the characteristics of the data space as in JMPEG, 18(3) 2009, JMPEG 20(9), 2011, (Lambrakos). Inverse models will represent the AM heat source and work piece by simple, path weighted diffusivity, providing for inclusion of all experimental and validated simulation data into the data space. Inverse functions are inherently structured for the representation or over determination by large data sets and are insensitive to strong or nonlinear transitions allowing the embedding of multi-scale temporal or spatial data imbedding fine time scale data into coarse time scale solutions.

Fabrication at point of urgent need will be enabled by airlifting an AM/SM fabrication machine as part of a mobile hospital into a war zone, disaster relief effort, or to limited access locations within conflict regions. Mobile hybrid fabrication units, such as on-board ships providing replacement part inventories for remote locations such as off-shore oil platforms will replace inventory spares, such as values and pumps, where long leads times and high transportation costs are a factor. Off the grid fabrication capability will be established using locally sourced materials and energy sources, not just for lunar environments.

Telepresence will offer the integration of human control and decision making with semi-autonomous remote operations. Human augmented robots capable of service and repair work using AM and other means may be dispatched to work in remote or highly hazardous locations to perform difficult or dangerous tasks with a high level of precision and autonomy.

AM process optimization algorithms will speed AM deposition, reduce defects, improve accuracy, compensate for shrinkage, improve surface finish and produce functionally graded materials. Optimization of huge numbers of parameters linked across a wide range of technologies (design, materials, process engineering, fabrication), will result in significant savings in material, energy, and time.

The engineered development of cellular and shell structures for asymmetric anisotropic engineered properties such as tensile and compression strength, thermal expansion, energy absorption, vibration damping, flow impedance, thermal transfer, will provide additional directional degrees of freedom to optimize and realize a wide range of design freedom and performance benefits. The development of designs to utilize functionally graded materials of metals, alloys, composites, and ceramics will continue.

Automated hybrid machines, specially designed to enhance process observability and monitoring will create AM test samples, heat treat, machined to shape, test, and collect materials property data in a continuous process with little or no human support. Data generated will be sent directly to databases. A distributed network of identical test centers will assist in group cross validation of results.

Automated self-configuring experiments will plan and collect data for open data bases and intelligent systems to support the design of new AM alloys to determine AM deposit properties as a function of build conditions to create new algorithms for control. Algorithms that can learn from and fuse data derived from designed experiments will help populate existing databases and fill the knowledge gaps left open by limited application experience.

Evolutionary designs created using genetic algorithms will optimize both functionality and aesthetics. Personal preference will fuse with design know how to create one-of-a-kind objects. As an example, if you were shown images of 200 classic sports car designs, asked to pick a few and then state your preference for performance and economy, an algorithm could design a car for you based upon both your emotional and technical preferences.

Knowledge capture and knowledge recycling applications may emerge as a new business model. As old tooling is scrapped from out of date factories, or sunset industries, tooling geometry is scanned and fed into the cloud, as are manuals, blueprints, other documents, or books (think project Gutenberg technology in 3D). This may be yet another function of existing, or newly formed, mobile knowledge recycling firms, capturing, and recycling industrial business knowledge, at the point of dismantling or scrapping. Knowledge is recaptured and recycled by knowledge salvage companies and sold to on demand AM fabricators or resellers. Big Data may be able to recognize and prioritize obsolete product streams, sunset companies without the knowledge or means to archive. Artificial intelligence (AI)I knowledge capture will identify and back up critical industrial systems needed to serve critical functions for many more decades and to support the sustained need for service, maintenance, and replacement parts. AI will create and "endangered infrastructure" list creating, saving and protecting critical infrastructure information.

Shared data, common databases and cloud computing will allow free access to serve the global need for designs. This will allow developing nations or under-served regions the opportunity to leap frog those countries and economies held back by the inertia of legacy infrastructure. The internet of things will be extended to piece parts and critical components ranging from bridges and buildings to pieces of art fabricated by AM with integrated sensors turning each part into a data gathering node.

Large databases for AM will be populated allowing AI systems to determine what works, what does not work and how to optimize and serve the needs for components across the globe. Prediction on what is needed, where in the world, will help drive local supply needs as well as serve global decision makers.

In Big Data: A Revolution that will transform how we live, work and think. (Mayer-Schönberger and Cukier 2013). The authors' state:

> Predictive analytics, what but not why, will be "good enough" to optimize based on very large data sets. UPS, 60,000 vehicle are monitored. Forensics and lessons learned... understanding what comes next based on correlation vs a first principals understanding. In some cases such as AM a first principals understanding of all the physics and materials properties may be out of reach but in the nearer term big data may offer a way to obtain functional mappings of the many process parameters and the outcome or quality of the part. Data driven decisions may trump first principal physics understanding where the big data geeks may win out over the subject matter experts and academics.

In Rise of the Robots, Martin Ford states (Ford 2015):

> IBM researchers confronted one of the primary tenets of the big data revolution: the idea that prediction based on correlation is sufficient and that a deep understanding of causation is usually both unachievable and unnecessary.

12.4   Business and Commerce

The business and commerce sector are taking notice as AM technology has progressed from demonstration to application. Rapid prototyping has proven itself as a viable application of the technology but the huge returns must come from industrial sector applications. Revenue trends for the total number of AM metal machines sold sale of AM metal feedstock and are rapidly increasing (Fig. 12.5a, b). As soon as businesses start making or saving significant money people take notice. As stated earlier in the book, boom and bust cycles of technology and the economy are a fact of life as no one has a crystal ball and prophets are only recognized when studying the past. The promise of rapid prototyping and AM technology has been slow in coming and the "killer applications," where returns greatly outweigh past

(a)

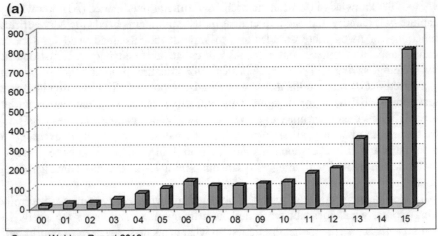

Source: Wohlers Report 2016

(b)

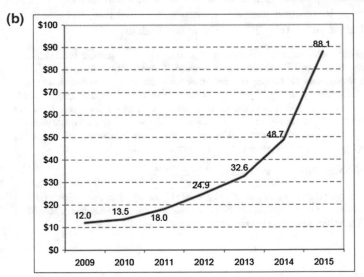

Source: Wohlers Report 2016

Fig. 12.5 a Number of AM Metal Machines Sold Per Year (2000–2015).[46] **b** Value of AM Metal Feedstock Sold in $M[47]

investments, are few and far between. Having said that, huge investments are being made and factories being built based upon a growing realization of the benefits and betting on the future returns of this technology (Fig. 12.6).

---

[46]Courtesy of Wohlers Report 2016, reproduced with permission.
[47]Courtesy of Wohlers Report 2016, reproduced with permission.

A series of McKinsey reports related to 3D printing benefits in *Remaking the Industrial Economy*[48] provide a professional assessment of how world manufacturing will change as a result of additive manufacturing and global economics. The current *linear economic model* of make, use, and dispose is contrasted with one of *circular economics* with feedback loops emphasizing *reuse, refurbish, remanufacture*, and the decoupling of economic growth from material intake and are projected for a future world of limited resources. In *Next-Shoring*,[49] the location of manufacturing proximity to demand and locations of innovation, is contrasted with the current *Off-Shoring* model as developing country wage and standard of living differentials decrease. They state 3 billion new consumers will enter the middle class by 2030. The current supply models and supply chains must change in a world of limited resources. The McKinsey report *Disruptive technologies: Advances that will transform life, business and the global economy*, referenced earlier in the book, is complementary reading to these reports specific to 3D printing.[50]

Changes to the value added stream will be realized through personalization, reduced time to market, reduced market entry costs, moving production closer to the point of consumption and hybridizing the conventional model of retail supply and demand, seamlessly fusing additive, and subtractive manufacturing. Product development cycles are already being altered by rapidly responding to changing demands, forecasts and inventories as described in a subsequent McKinsey Global Initiatives report.[51]

AM will speed product development with rapid prototypes, speed and test the adoption curve and minimize risk using small lot production tooling. In addition, further speed to market will be realized by AM efficiently creating complex master patterns, jigs, fixtures, and manufacturing aids, while more efficiently managing the "market tail" transitioning to warehousing virtual designs and tooling, reducing obsolete inventory and designing functional replacement parts for single or small lot AM production.

In one example, Senvol[52] offers a free database for searching machines and materials and provides analytical services to help users and potential users of AM technology to select machines, materials, service bureaus, and target applications

---

[48]McKinsey & Company report, "3-D printing takes shape", January 2014, by Daniel Cohen, Matthew Sargeant, Ken Somers, http://www.mckinsey.com/insights/manufacturing/3-d_printing_takes_shape, (accessed March 27, 2015).

[49]McKinsey & Company report, "Next Shoring a CEO's Guide", January 2014, Katy George, Sree Ramaswamy, Lou Rassey, http://www.mckinsey.com/insights/manufacturing/next-shoring_a_ceos_guide, (accessed March 27, 2015).

[50]McKinsey Global Initiative, "Disruptive technologies: Advances that will transform life, business and the global economy", May 2013, James Manyika, Michael Chui, Jacques Bughin, Richard Dobbs, Peter Bisson, Alex Marrs, http://www.mckinsey.com/insights/business_technology/disruptive_technologies, (accessed March 27, 2015).

[51]McKinsey & Company report, "Are you ready for 3D Printing?" February 2015, Daniel Cohen, Katy George, Colin Shaw, http://www.mckinsey.com/insights/manufacturing/are_you_ready_for_3-d_printing?cid=other-eml-alt-mkq-mck-oth-1502, (accessed March 27, 2015).

[52]Senvol free database, http://senvol.com/database/, (accessed April 6, 2015).

and do cost, market, and supply chain analysis. To date, the proprietary algorithm uses 30 search fields to assist the customer. Viable outcomes will be hard to identify, such as service provider costs vs. quality or cost savings, across a nonexistent or hypothetical AM supply chain. Disruption across the supply chain in the cost of materials, energy, and computing and information technology will make the viability of any predictions and analysis suspect.

Private membership databases and Web sites may be established to allow uploading, sharing, and downloading of proprietary designs, design information or process information, used to realize or validate these designs. Commercial vendors may sponsor such databases with formal contribution rules and guidelines to direct or help control the quality of the contribution. Design information may range widely from unique applications to controlled test conditions, in service reports, or formal failure analysis. Contributors may be assigned "Member Credits" or "Member Ratings" accrued by the number and quality of content contribution. Member Ratings could then be used to help validate or support other business interests. Private sector databases may use other options to encourage submission such as direct payment or payment in credits for equipment, supplies, or services. Submissions leading to intellectual property generation may be worthy of micropayments for the life of the patent.

In the book *Producing Prosperity* by Gary P. Pisano and Willy C. Shih (Pisano 2012):

> Manufacturing has become knowledge work. Manufacturing is often highly integrated into the innovation process.

> Knowledge intellectual property and information can flow across sectors, spurring innovations across sectors. Innovation in one sector of industry can flow across sectors can never really be viewed as isolated from what happens in other sectors.

In *Unleashing the Second American Century: Four Forces for Economic Dominance* (Kurtzman 2014), Joel Kurtzman, a senior fellow at the Milken Institute, identifies four transformational economic forces: unrivaled manufacturing depth, soaring levels of creativity, massive new energy sources, and gigantic amounts of capital waiting to be invested. He forecasts a bright and strong future for the American economy and describes the fastest growing economic sectors as bio-tech, pharmaceuticals, computer hardware, telecommunications, advanced manufacturing, materials science, aeronautical, and space engineering. He specifically calls out the potential of 3D printing which could play a role in many of these sectors. This book indicates an awareness of 3D printing technology's role in the future of American manufacturing at the highest levels of government, and academia.

In *Zero to One*, Peter Thiel (Thiel 2015) discusses the value of creating something entirely new rather than just doing more of what everyone else is already doing: in essence, taking an idea and a business from 0 to 1. He discusses the creation and protection of proprietary technology, and how to fly under the radar to avoid competition while radically improving a product, service, or an existing solution by a factor of 10. Superior designs and products can result from the integration of multiple new technologies. What important truth do very few people agree with? What valuable company or product is nobody building? What undiscovered aspects of the physical world can be realized? While interpreting this book within the context of AM, one can restate those questions and draw upon your own knowledge, experience and the possibilities offered by this new technology. Perhaps that will inspire you to set up your own AM skunk works in the back shop space.

Many companies are getting into the game but telling us more of what they are planning to do rather than what they are doing or have done. It is not uncommon to see a press release or article chock full of forward-looking statements identified by words, such as "anticipates," "intends," "plans," "seeks," "believes," "estimates," "expects," and other similar references to future events. Should the buyers beware? We will see.

Consumer Apps for mobile or hand-held devices will emerge allowing highly customized and individualized specification and purchase of consumer goods and custom built products where and when they are desired. Designs will be available on the Web or on mobile devices. Individualized souvenirs could be one example. They will be simple to use and may be offered as customized by personal selection or as your historical preference, search and purchase data indicates.

The value of a design may be related to its information content, popularity of download, popular appeal, or validated usage in industry. Individual designers can already post designs for sale and make them available for 3D printing. Group source designs allow participation and reward for contribution to the design process, with micropayments extending through the service life of the component.

Crowd thinking and innovation can be harnessed through connectivity. Given the right incentives, widely dispersed users, inventors, engineers, and makers can rapidly be brought together to speed products to market. The Web site *Quirky* is often cited as one such example.

Consumer specific data, as provided by consumers to open source data brokers, will be available to offer discounts and custom deals made to your specifications and preferences in accordance with current styles and trends. Algorithms will be able to design the virtual product and market it directly to you at a price that not only reflects the cost of the product but factors in production capacity and other variables of the market, such as market penetration. As an example, if they know you are a golfer and they know you like high end toys and the golf club AM fabrication machine is idle, they may cut you a huge discount to lure you to buy the latest putter design to show off at your country club.

Algorithms for B2B (business to business) and M2 M (machine to machine) communication will seamlessly integrate the design and sales through manufacturing and delivery. Fabrication of the part will proceed and be completed without the need for human intervention.

Specialized "Business and Engineering Apps" will continue to be developed to infuse the power of seamlessly integrated CAD/CAE/FEA/CAM/AM/SM, material selection, process definition into low cost but high powered menu-driven solutions for specialty applications such as medical devices, hearing aids, automotive and aerospace sectors. This will open up this capability to a wider range of small to medium companies that cannot support a full time staff of engineering experts in all of the required AM skills.

Specialized "Consumer Sites" will offer custom consumer products (e.g., sports equipment.) and virtual warehouse of goods that can be customized to consumer specification and choices. A consumer may select styles, colors and customized shapes and sizes of goods tailored to them (one size fits one) and expect next day delivery. As a direct affront to privacy concerns consumers may be able to provide secure links to personal data to include full body scans and other personal data that may be shared with retailers and other providers. Scan stations and scan services will emerge, where consumers may have themselves or items of their choice scanned to allow personalization of designs.

Networks of small companies, resources, and services in the manufacturingchain will be linked to offer local capability in design, fabrication, and finishing. User groups, such as the Additive Manufacturing Users Group (AMUG) will continue to provide an annual forum for networking and education to advance the new and commercial uses and applications of AM technology. Independent training resources are being established to help bridge the gap between the AM industry and users to promote awareness, safety, quality and innovation.[53]

Large corporations and investors are gaining interest and attempting to identify markets and trends. Business models will need to be redeveloped to accommodate and best utilize the cradle to grave supply chain of goods and services affected by this new paradigm. Business and consultancy firms are being established to educate and help guide companies to identify and take advantage of the opportunities offered by AM technology.[54]

An article from the Techrepublic Web site, "3D printing: 10 companies using it in ground-breaking ways[55]" by Lyndsey Gilpin, reports that a growing number of innovative companies experimenting with 3D printers, is propelling the technology closer to the mainstream market.

Governments will need to develop answers to questions such as, how do workers transition to other jobs or how to harness the world population of new workers to gainfully contribute to society? How can AM fit into this big picture? Concerns regarding a negative impact upon the workforce are always raised upon the introduction of a new technology. The example often cited is that of the Luddites, textile artisans of the early 1800's violently protesting the introduction of new textile technology and machinery, fearing job loss. To an extent this happens in all new or optimized industrial processes or with almost any work force where the old

---

[53]UL Additive Manufacturing Web site, http://industries.ul.com/additive-manufacturing, (accessed January 3, 2017).

[54]Deloitte Web site for manufacturing, https://www2.deloitte.com/us/en/industries/manufacturing.html?icid=top_manufacturing, (accessed January 3, 2017).

[55]"3D printing: 10 companies using it in ground-breaking ways", by Lyndsey Gilpin, March 26, 2014, http://www.techrepublic.com/article/3d-printing-10-companies-using-it-in-ground-breaking-ways/, (accessed April 12, 2015).

skills are replaced with the need for new skills. AM is not alone in the need to establish a pipeline for learning and retraining to develop the skills needed for advanced manufacturing. The question of how this is best done remains open.

A recent report by LUX Research, "Building the future: Assessing 3D printing's opportunities and challenges,"[56] assesses the markets and business opportunities for AM. They see market development led by automotive, medical and aerospace but also a future referred to a reshaping of the "manufacturing ecosystem."

12.5 Intellectual Property, Security, and Regulation

Information drives the global economy. It is an increasing part of all of our lives and our dependence upon it is growing, perhaps without bounds. Information is a big component of knowledge and knowledge is a big component of power. AM has been described as a means for personal empowerment, with terms, such as *democratization* and *independence* often being used. Information has value. The value of creating, accessing, and protecting information, increases as the ratio of information associated with our livelihood, wellbeing, and standard of living increase.

Patent, copyright, trademark, and other legal issues will force a rapid evolution and application of laws, such as the Digital Millennium Copyright Act (DMCA) and the Stop Online Piracy Act. Legal, security and regulatory issues are listed in Fig. 12.6.

Product or design liability may become an issue if you post a unique design on the internet, then receive a micropayment from someone who builds it and dies using it. What is your legal exposure? How do you ensure that the part you build for someone else (e.g., motorcycle wheel) will perform the same as a design posted on the internet? Certification of AM products, designs, materials, and processes will be a complex process relying on a huge body of quality and currently nonexistent, material and process data.

Copyright and patent law has some catching up to do regarding near term development of 3D printing and AM. In general, copyright covers art and expression, while patents cover works of utility but if you scan a shape and recreate it out of the same material, is there an infringement? How different must be the materials, shape and process, to constitute an infringement? If you produce a unique design, process, material, or end product how do you protect it? How do you share it? What means are available to benefit from its commercial value and use? The legal system as well as state and federal laws has some catching up to do.

Global patent and copyright laws vary from country to country with a wide range of enforcement. The music and movie industries have demonstrated copyright enforcement is difficult at best. Global intellectual property must be realized and

---

[56]LUX Research, "Building the future: Assessing 3D printing's opportunities and challenges," https://portal.luxresearchinc.com/research/report_excerpt/13277, (accessed March 27, 2015).

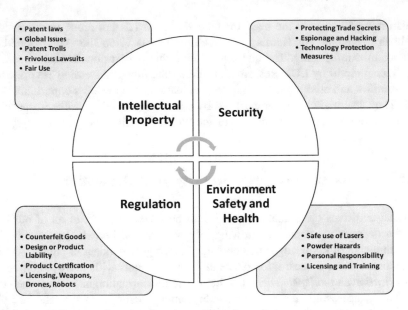

Fig. 12.6 Additive Manufacturing Security and Regulatory Issues

protected perhaps by the development of ©-stamped designs, ©-stamped codes, or ©-stamped content.

Regulation of both AM system hardware and design data will emerge. As lasers become smaller, cheaper and more powerful, such as used in the 3D printing of metal, regulation will be needed to place greater demands on controlling their usage. It is one thing to shine a hand-held Class I laser pointer at a place kicker at a football game. It is another to misuse a Class IV laser from an AM machine. Is the legal system and law enforcement prepared to locate and confiscate Class IV lasers that are being used improperly or hazardously?

The detection of counterfeit consumer goods or industrial components fabricated using inferior materials or noncertified processes will become necessary. This problem will be akin to that of counterfeit nuts and bolts but may extend to more complex components used in critical applications. Legal and technical means to detect, identify and track inferior materials and goods across complex supply chains will need to be developed to ensure safety and performance to the end user but also to protect companies and brand names from damage.

A Wired magazine article by Clive Thompson, "3-D Printing's Legal Morass,[57]" refers to the Digital Millennium Copyright Act and Stop Online Piracy Act, as 3D printing will create challenges to the copying of physical objects and enforcement as 3D printers incorporate scanners and become 3D copiers. In the article he states

---

[57]"3-D Printing's Legal Morass", Wired, 05.30.12, by Clive Thompson, http://www.wired.com/2012/05/3-d-printing-patent-law/, (accessed March 28, 2015).

"patents typically cover the function of objects and expire after 20 years, but not piece or replacement parts, while copying artistic patterns or designs on an object are protected under copyright law". Also in the whitepaper "What's the deal with copyright and online printing?[58]" Michael Weinberg does a god job of discussing what is patentable and what may or may not be copyrighted, along with legal concepts of severability and merger as it applies to creative or functional works. While scans may not be sufficiently original to be copyrightable, a useful object may be patentable. To be patentable it must be useful, novel and nonobvious. The article uses an example that a CAD file may be copyrighted, but a 3D printout of the useful object is not. Additional detail is provided by Public Knowledge in the whitepaper by Michael Weinberg, "3 Steps for Licensing Your 3D Printed Stuff.[59]"

Trade secret law is evolving in an attempt to keep up with information, privacy, cyber security, hacking and a highly mobile, global workforce. The U.S. government has studies the problem,[60] developed a strategy and is enacting laws, such as the Defend Trade Secrets Act (DTSA) of 2016,[61] to mitigate the problem. Trade secrets are comprised of three fundamental components: (1) information not generally known in the industry, (2) measures taken to protect the information, and (3) the economic value of information kept secret. In the article, "The Other IP: Trade Secret Law and 3D Printing,"[62] Bryan J. Vogel states components of the law include, trade secret law does not require filings or government approval, to offer protection to smaller less-well-funded innovators to protect their IP to sustain a competitive advantage.

Industrial espionage will increase as will the efforts and methods used to counter these threats. Arrests are already being made as described in the article "*3D Printing and Industrial Espionage: Former United Technologies Corp. employee arrested,*[63]" by TE Edwards. The article goes on to say the InfoSec Institute quotes,

---

[58]Institute for Emerging Innovation, "What's the Deal with Copyright and 3D printing?" Michael Weinberg, January 2013, https://www.publicknowledge.org/files/What.%27s%20the%20Deal%20with%20Copyright_%20Final%20version2.pdf, (accessed March 28, 2015).

[59]Public Knowledge, "3 Steps for Licensing Your 3D Printed Stuff," March 6, 2015, Michael Weinberg, https://www.publicknowledge.org/documents/3-steps-for-licensing-your-3d-printed-stuff, (accessed March 28, 2015).

[60]Administration Strategy on Mitigating the Theft of Trade Secrets, Executive Office of the President of the United States, February 2013, https://www.whitehouse.gov/sites/default/files/omb/IPEC/admin_strategy_on_mitigating_the_theft_of_u.s._trade_secrets.pdf, (accessed May 15, 2016).

[61]Protections of Defend Trade Secrets Act article, The Nation Law Review, May 12, 2016, http://www.natlawreview.com/article/protections-newly-enacted-defend-trade-secrets-act, (accessed May 16, 2016).

[62]Inside3dp.com article, "The Other IP: Trade Secret Law and 3D Printing", By Bryan J. Vogel, on Dec 3 2014, http://www.inside3dp.com/ip-trade-secret-law-3d-printing/, (accessed March 28, 2015).

[63]3dprint.com, 3D Printing and Industrial Espionage: Former United Technologies Corp. employee arrested, by TE Edwards, December 11, 2014, http://3dprint.com/30297/yu-long-industrial-espionage/, (accessed March 28, 2015).

"Counterfeit products could be introduced into the supply chain in order to cause anomalies and faults in the products designed."

The US government is well aware of the potential risks to a digital manufacturing environment and is trying to stay ahead, as evidenced by the NIST Cybersecurity for DDM Direct Digital Manufacturing symposium[64] which covers attack scenarios, challenges, gaps, other potential scenarios, the reach of threats, and protecting technical data on the factory floor.

Digital corporate espionage will force tighter and stricter controls on information, imposing embargos on a wider array of enabling technologies. Virtual states may emerge in a distributed or underground fashion with their own citizens, economies, and e-national interests. Sabotage, infiltration and corruption of digital design or build hardware will need to be confronted, detected, and protected against.

Global security issues associated with additive manufacturing will emerge as ordinary citizens, empowered by wireless internet cloud access to information, will be able to fabricate complex objects. Drone or robotic structures may be fabricated and assembled in low profile facilities with a minimum of smuggled components and a minimum of equipment in covert hidden locations, as boots on the ground are replaced with bots on the ground (Schmidt 2013, p. 153).

Frivolous lawsuit claims of infringement and false advertising may also come into play. If someone is asserting copyright and claiming infringement, this may not always be the case. In cases such as these, a bit of research or advice from the user community or legal means, may offer a different opinion. Do not blindly assume copyright covers everything. The integrity of information may also come into play. The assertion or advertisement of patent protection does not mean the patent is in force. The USPTO offers tools and interfaces to allow online patent searches to identify patents pending, those granted, those that have expired due to failure to pay the appropriate patent maintenance fees and those that have been canceled by the U.S. Patent and Trademark Office.

The definition and detrimental roll of the *patent troll* has come into the spotlight as the rapid pace of technology development has often out stripped the legal means to protect it. Patent trolls buy or misuse patents in an effort to launch lawsuits or litigate as a business strategy for profit, undermining the intent to protect the inventor or intellectual property owner. They can throw up roadblocks to freeze or prevent other companies from practicing or developing the technology.

All one needs to do is follow a Google Scholar alert feed for "additive manufacturing" for a few months to realize the frequency of patent issues related to the technology. How many of these will ultimately benefit and protect the inventor and how many will end up in the hands of trolls? John Hornick an intellectual property

---

[64]NIST Cybersecurity for DDM Direct Digital Manufacturing symposium, Gaithersburg, MD, February 3, 2015, agenda, http://www.nist.gov/itl/csd/cybersecurity-for-direct-digital-manufacturing-symposium.cfm, (accessed March 28, 2015).

(IP) attorney specializing in 3D Printing law and author (Hornick 2015), discusses the challenges that 3D printing will make to the future of IP in his book and the video "3D Printing and the Future (or Demise) of Intellectual Property."[65]

Components that could result in loss of life or critical function, such as automotive, medical, or aerospace are subject to regulation by various governing bodies. This applies for all conventional processing as well as AM components. The greater risk associated with AM is the fabrication of unregulated components made easier with AM and less easy to detect.

Technology protection measures (TPM) can be put into place by manufacturers to prevent users of the printer from accepting nonmanufacturer approved feedstock. Limitations to use imposed by TPM may be challenged under law as corporate interests or government policies can be at odds with public interests. Issues associated with the development and applications of AM include technology innovation, public domain, and fair use. Public Knowledge[66] is an organization with a stated mission to preserve the openness of the Internet and the public's access to knowledge. It promotes creativity through balanced copyright, upholds. and protects the rights of consumers to use innovative technology lawfully and advocates for policies that serve the public interest. A petition, for exemption of 3D printer users from TPM for 3D printing feedstock controls has recently been recently granted.[67]

A trend for increasing data security is needed to protect intellectual property of the model or build sequence. Certified secure fabrication centers may be created and rely upon secure data storage, data transfer, cloud links, and hardened decryption chips to decode and protect encrypted design and process data in-real-time, at-the-machine, to help ensure the protection and monetary value of this information. EPROM (erasable programmable read only memory) encryption will create and separate file encryption and hardware encryption codes only to be combined and unlocked during the fabrication event. One option, data annihilation, will erase any trace of information transmitted or generated during the build sequence. Another option would be to pass along an ownership key to prevent data destruction and allow subsequent build or marketing of this information. These fabrication centers may operate in real time 24/7 with resource availability being known to cloud clients from anywhere in now-time. Secure design to part protection can reduce potential liability of a design or processing stream from being hacked.

[65]John Hornick, YouTube video, "3D Printing and the Future (or Demise) of Intellectual Property" http://www.youtube.com/watch?v=JoIjUKlwFkA&feature=c4-overview (accessed March 23, 2016).

[66]Public Knowledge web site, http://www.publicknowledge.org/about-us/, (accessed March 28, 2015).

[67]Docket No. 2014-07, http://copyright.gov/1201/2015/fedreg-publicinspectionFR.pdf, (accessed February 1, 2016).

As an example, Authentise[68] offers business to business (B2B) cloud software to enable secure 3D printing. A wide range of security measures to protect IP (intellectual property), design integrity and the data transmission process have been developed to print designs securely. They offer a wide range of technology to stream data to printers securely and provide other security measures. As an example, products such as the Authentise Computer Vision Web cam can provide build time monitoring and interactive mobile control to mount and interface with a 3D printer.

Protection of intellectual property must extend across the globe and multiple jurisdictions as they currently lack harmonization. The enactment of laws, both domestic and foreign, to offer such protections, are likely to lag technology development as business practices typically lag the adoption of new technology. As the pace of technology development quickens, the pace at which businesses adapt to these changes must also evolve. Companies are in a race with each other to assess the benefits of AM to their current and future product lines, retrain, and reinvent their design know-how.[69] Just as workers will need to continually evolve their IP protection knowledge and skills and so must companies and corporate structures.

Environment, safety, and health (ES&H) issues may need additional regulation and tracking due to the democratization of manufacturing that relies on the use of high energy beams and potentially explosive or toxic powders. Unlike a 3D plastic printer you may set up in your office or garage, metal melting systems, and systems utilizing metal powders will pose many greater risks. Nonprofessional and unlicensed use and access may require a higher degree of training and compliance to operate safely and appropriately.

Embedded security using quantum dot technology is being developed in a collaboration between Quantum Materials Corporation and the Institute for Critical Technology and Applied Science and the Design, Research and Education for Additive Manufacturing Systems (DREAMS) Laboratory at Virginia Tech. A Quantum Materials Web site post[70] describes their quantum dot security technology that can embed quantum dots within objects being 3D printed to produce a physical unclonable signature to help ensure positive identity of an object.

Finally, economic security comes into play when certain organizations have the job of worrying about the improbable or unlikely. As an example, the US Department of Homeland security asks if all the IT, Web, and electric power grid went away so will AM, what then? While it is unlikely that conventional manufacturing will disappear anytime soon, it is valid to ask whether AM will ever become a national resource and a national security risk?

---

[68] Authentise Web site, http://authentise.com/, (accessed March 28, 2015).

[69] McKinsey & Company report, "Next Shoring a CEO's Guide," January 2014, Katy George, Sree Ramaswamy, Lou Rassey, http://www.mckinsey.com/insights/manufacturing/next-shoring_a_ceos_guide, (accessed March 27, 2015).

[70] Quantum Materials web site post, http://www.qmcdots.com/products/products-3dprinting.php, (accessed March 28, 2015).

12.6 Social and Global Trends

As throughout history, technology has been created to serve society and individuals, sometimes for personal enhancement, sometimes for the good of all, sometimes to gain advantage or to be used against one another.

Global Trends 2030: Alternative Worlds provides a framework for thinking about the future social and global trends, possible scenarios, and potential outcomes.[71]

The framework offered by this report breaks down into four mega trends:

- Individual empowerment, poverty reduction, growing middle class, greater education widespread communications, and healthcare advances
- Diffusion of power, networks, and coalitions, multi-polar world
- Demographic patterns, aging, urbanization, migration
- Food, water and energy, population growth, supply, and demand.

We find this framework useful and will discuss the potential impact of 3DMP and AM within each category (Fig. 12.7).

Individual Empowerment

As stated early in the forward to the book: metal is valuable, it empowers people with objects, it enables people to display and project power. Widespread communications, Web access, and advances to information and technology will raise the education levels, reduce poverty, and grow and redefine a new global middle class. The democratization of manufacturing offered by access to AM technology will allow people access to personalized objects on demand, where needed and when needed. This will increase the utility, productivity and relevance of people across the globe. People will be empowered by better health care and longer, more productive lives. People will have access to personal drones and robots, extending the capability, and reach of their personal power. Additive manufacturing will play a part in all of these technologies.

Products in support of locally manufactured goods and healthcare products will be available at low costs in places previously unavailable. Local or perhaps even mobile manufacturing cells will be set up near energy sources (water, wind, solar, fossil fuels, burning garbage, or methane) or sources of raw materials (metal, glass, plastics), such as recycle stations. Satellite and drone-based internet, cloud, and Web access will provide access to open-source models for a wide range of consumer goods, construction hardware, and infrastructure support. Access to Web-based education will allow remote learning and employment opportunities

---

[71]National Intelligence Council, Global Trends 2030: Alternative Worlds provides a framework for thinking about the future social and global trends, possible scenarios, and potential outcomes, December 2012, NIC 2012-001, ISBN 978-1-929667-21-5, http://www.dni.gov/files/documents/GlobalTrends_2030.pdf, (accessed March 28, 2015).

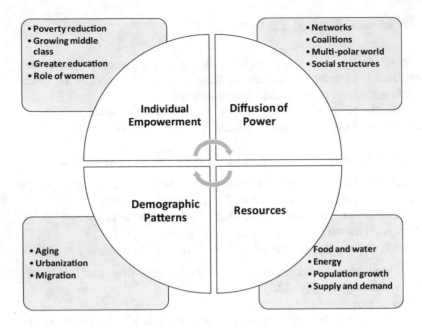

Fig. 12.7 Global Mega Trends Impacting 3D Printing and Additive Manufacturing

associated with producing AM models for use across the globe, modifying or customizing existing models for specialized or individualized applications.

Goods specifically tailored for one's geographic, demographic, and individual wants and needs will be available in an increasingly virtual human environment. Human nature will still want things to have, hold and possess serving the empowerment of ownership. All goods will be increasingly data driven. Many will be produced beyond control or oversight.

Affluent populations will have access to high quality unique designs made to individual specifications such as premium health care and quality of life hardware. Exoskeletons, implants and human augmentation devices will rely on a fully described set of your personal data down to your DNA level. A full data set of your body and health specifications, such as bone structure, age, and health will be used in manufacturing custom devices. Clothing to fit and customer goods, optimized for use, will be based upon stated or predicted needs.

How will individual empowerment affect views of religion and appeal to basic human nature? This will likely be defined by each of us. Individual empowerment will lead to an increasing role for women within all aspects of society, politics, and world order.

12.6.1 Diffusion of Power

The rise in individual empowerment may coincide with a decrease in governmental power, resulting in an increase in personal freedom, and the necessity of people to associate and align with non-governmental agencies. Corporations and networks of likeminded social or professional groupings will emerge to fill the gap if countries fail to overcome their differences and work together for a common good. Corporations may join together to provide benefits to employees and their families, offering solutions for healthcare, transportation, housing, security, information, social, and quality of life, not provided by national, state or city governments. Social stratification may result. Professional villages and virtual communities may emerge as hybrid social units offering a group plan-based alternative community structure, allowing a demographic to opt out of government restriction or regulations. Nonaligned social structures will increasingly rely on the generation of resources and energy and the vertical integration of materials, goods, and services such as provided by AM.

Third world ascendance will be empowered by localized energy sources, food creation, and manufacturing capability, bypassing the need to access the historical information or constraints of historical manufacturing infrastructure. How will the militarization of space affect the balance of power? Will corporate access to space evolve into personal access to space? A reference to the changing power structure envisioned for the next 30 years is given in (Naím 2013) *The End of Power: From Boardrooms to Battlefields and Churches to States, Why Being in Charge is not what it used to be.*

12.6.2 Demographics, Information, Mobility, Education, Connectivity

Radical adoption of AM technology will occur in highly underserved countries, as in the example of Finland where little infrastructure existed in land line telephone service. Widespread investment and adoption of the new cell phone (Nokia) technology was expedited and enabled both by the push of newly available technology and pull of technology need.

While people become more mobile they will have less reason to travel as all the basic necessities will be provided at their door step. Today, we have over 7 billion people on the planet; by 2030, it is estimated we will have over 8 billion people. This will cause us to rethink what we need and how to use the available energy, materials, and resources most efficiently. We will need increased efficiency of all systems. How will AM play into this? Individuals will own fewer cars and travel less as goods and employment are delivered right into the home as required. Schedule and cost share of resources will reduce the cost of maintaining a relatively high standard of living. The global definition of the good life, once referred to as the

American Dream, has flattened and become more information-driven and focused or meeting individual needs in an increasingly virtual manner. Shared objects, or services, such as autonomous taxis and delivery trucks will allow extending the delivery hours without violating labor laws. An aging population will affect consumer products, mobility, robots, exoskeletons, personal robots, and drone delivery.

Flattening of the global standard of living will inevitability result as 8–9 billion people will not be able to live like today's developed world population. Huge improvements will occur in the provision of global basic needs, such as food, water, energy, and education but access, quality, quantity, or choice of other services will change significantly. Health care, as one example, may be free to all at the price of providing Big Data with your DNA, skeletal 3D definition, and personal activity imprint. 3D skeletal data may be required as part of an 80 year tune up, replacing all worn out joints, teeth or weak bones in one operation. Standardization and limitations of options may be another cost. While the world flattens it will also grow smaller. The concepts of an extended common space will include cyberspace, outer space, marine, and airspace.

The information ratio and virtual experiences associated with our lives has risen significantly over the past decades. Information about who we are, what we are and predictions of what we can be will be updated by our every action. In most cases the virtual experience, with its personal information-driven focus, will exceed much of what we could provide for ourselves in the physical world.

12.6.3   Food, Water, Energy, Population Growth

The shift and migration of population to the mega cities will more efficiently accommodate the burgeoning human population. Vertical urban greenhouses and the production of artificial food will localize and optimize production and supply. New means to harvest and produce energy, and to recycle materials and water have evolved to support population growth. Excesses in supply and demand, production, and consumption will narrow. Resources and consumption will decrease and be minimized.

12.7   Trends in Additive Manufacturing

As we have seen in this chapter, it is easy to spot trends in AM technology development and make predictions of where it may go over the next five to ten years. But it is certainly harder to put it all on a schedule, know where to invest time and money, where and when to expect monetary returns and how to weigh the opportunity costs of taking the plunge into AM metal processing. The landscape of AM technology and companies is constantly changing with companies being launched, divided, and acquired.

12.7.1 Top AM Technology and Market Destinations

I will end this last chapter with my top picks for the technology destinations and market destinations (Fig. 12.8) providing the greatest impact for AM metal:

(1) Lower cost, higher quality materials and metal alloys will propel high-performance alloys into greater use.
(2) Software and fully integrated parametric model-based design and engineering analysis from CAD through AM fabrication with universal translation of CAD file formats and across AM platforms.
(3) Automated material property generating systems gathering data to populate databases, supply predictive models, and will assist in the certification of materials, processes, and components. Improve models to help understand the impact of complex geometries on material properties and part performance.
(4) Open architectures for AM systems enabling in-process data gathering and links to part inspection data with feedback into design and process simulation and for populating databases and driving predictive models will proliferate.
(5) Improved capacity, speed, and accuracy of existing AM technology reducing failure and scrap rates seen during AM fabrication will help drive adoption.
(6) Hybrid integration of AM with SM, on-machine and in-process inspection.
(7) Greater access and lower costs provided by service providers' will greatly expanding rapid metal prototyping into the small-and mid-size business sector and to individuals.
(8) Advanced low cost, high accuracy 3D scanning tools allowing interface with real-world geometry, reverse engineering and full integration into the parametric model-based engineering environment.

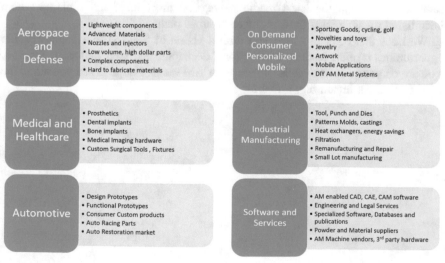

AM Metal Roadmap Top Destinations

Aerospace and Defense
- Lightweight components
- Advanced Materials
- Nozzles and injectors
- Low volume, high dollar parts
- Complex components
- Hard to fabricate materials

On Demand Consumer Personalized Mobile
- Sporting Goods, cycling, golf
- Novelties and toys
- Jewelry
- Artwork
- Mobile Applications
- DIY AM Metal Systems

Medical and Healthcare
- Prosthetics
- Dental implants
- Bone implants
- Medical Imaging hardware
- Custom Surgical Tools, Fixtures

Industrial Manufacturing
- Tool, Punch and Dies
- Patterns Molds, castings
- Heat exchangers, energy savings
- Filtration
- Remanufacturing and Repair
- Small Lot manufacturing

Automotive
- Design Prototypes
- Functional Prototypes
- Consumer Custom products
- Auto Racing Parts
- Auto Restoration market

Software and Services
- AM enabled CAD, CAE, CAM software
- Engineering and Legal Services
- Specialized Software, Databases and publications
- Powder and Material suppliers
- AM Machine vendors, 3rd party hardware

Fig. 12.8 AM Technology and Market Destinations

(9) A wide expansion of training and STEM educational opportunities is needed, at the college and university levels but also at the community college, trade school, high school, and grade school levels.

(10) The continued generation of international standards to better integrating AM technology into the value stream of the global manufacturing economy.

12.7.2 The First Steps Toward AM Metal

This book introduced you to the language of AM metal, rolled out a map identifying destinations, the vehicles to take to the main attractions, hubs of activity, centers of learning and industry, highways, and byways. Seeing past the billboards of hype we focused on the street signs and learning the rules and language of the AM road. We have identified where you get on and off the conventional manufacturing highway and where to explore the points of interests and spend time with a view of a new and rapidly changing technology. We linked to the latest information showing where the bridges to other related technologies are being built, what roads are under construction, being improved and where the potholes are being filled. We mapped where the intersection of technologies will create growth and new areas of development. We provided a solid foundation of information needed to help you understand the financial investment and the skills needed to master the complexities of these AM processes and to develop realistic expectations of where this technology can take you or your business.

By finishing this book, you have already started the process of jumpstarting your knowledge and skills taking the first steps to putting AM metal technology to work for you. Following the Web links provided enables access to additional reading, instructive videos, sources of information and an introduction to the content of a full range of vendors, machine and service providers, governing organizations, centers of learning, the latest media news, and most importantly places where likeminded dreamers and thinkers meet or cross paths.

We hope this book provided you with a vision of the AM metal industry, the opportunities it offers leading to directions, decisions, and your first steps forward. You are on your way to explore the world of AM metal. I hope to see you along the way. Until then, dream big, travel far.

Professional Society and Organization Links

ACGIH, American Conference of Governmental Industry Hygienists,www.acgih.org
AMT, Association for Manufacturing Technology,http://www.amtonline.org/
ANSI, American National Standards Institute,www.ansi.org/
ASM International, (formerly American Society for Metals),http://www.asminternational.org/
ASME, American Society of Mechanical Engineers,https://www.asme.org/
ASNT, American Society for Non-Destructive Testing,https://www.asnt.org
ASTM International, formerly American Society for Testing and Materials,www.astm.org/
AWS, American Welding Society,http://www.aws.org
CISRO, Commonwealth Scientific and Industrial Research Organization,http://www.csiro.au
CSA Group, formerly Canadian Standard Organization,http://www.csagroup.org
DARPA, Defense Advanced Research Projects Agency,www.darpa.mil
EWI, Edison Welding Institute,http://ewi.org/
ISO, International Organization for Standardization,http://www.iso.org/iso/home
LIA, Laser Institute of America,http://www.lia.org
MPIF, Metal Powder Industries Federation,www.mpif.org
MIBP, Manufacturing and Industrial Base Policy office,http://www.acq.osd.mil/mibp
NIOSH, National Institute for Occupational Safety and Health,http://www.cdc.gov/niosh
OSHA, Occupational Safety and Health Organization,www.osha.gov
SAE International, Society of Automotive Engineers,www.sae.org
SME, Society of Manufacturing Engineers,https://www.sme.org
MS, The Minerals Metals and Materials Societyhttp://www.tms.org
USPTO, United States Patent and Trademark Office,www.uspto.gov
WIPO, World Intellectual Property Organization,www.wipo.int
WTO, World Trade Association,www.wto.org

© Springer International Publishing AG 2017
J.O. Milewski, *Additive Manufacturing of Metals*, Springer Series
in Materials Science 258, DOI 10.1007/978-3-319-58205-4

AM Machine and Service Resource Links

Welding Equipment and Consumables Suppliers

ESAB Group,http://www.esabna.com/
Fronius,http://www.fronius.com/cps/rde/xchg/fronius_usa
Lincoln Electric Company,http://www.lincolnelectric.com/en-us/Pages/default.aspx
Miller,http://www.millerwelds.com/

Additive Manufacturing Machine Builders and Service Providers

3D Systems,http://www.3dsystems.com/3d-printers/production/overview
Arcam AB,http://www.arcam.com/
BeAM Be Additive Manufacturing,http://www.beam-machines.fr/uk/en/
Concept laser,http://www.concept-laser.de/en/home.html
DMD3D Technology,http://www.dm3dtech.com/
DMLS.com,http://dmls.com/home
DMG Mori,http://us.dmgmori.com/dmg-mori-usa
EOS,http://www.eos.info/en
ExOne,http://www.exone.com/
Fabrisonic,http://fabrisonic.com/
Hybrid Manufacturing Technologies,http://www.hybridmanutech.com/
Liner Mold & Engineering,http://www.linearmold.com/
Materialise, software and services for 3D Printing and AM,http://www.materialise.com/
Matsuura Machinery Corporation,http://www.matsuura.co.jp/english/
MTI Albany,http://www.mtialbany.com/
Optomec Inc.,http://www.optomec.com/

© Springer International Publishing AG 2017
J.O. Milewski, *Additive Manufacturing of Metals*, Springer Series
in Materials Science 258, DOI 10.1007/978-3-319-58205-4

Realizer,http://www.realizer.com/
Renishaw, http://www.renishaw.com/en/laser-melting-systems–15240
RPM Innovations,http://www.rpm-innovations.com/laser_deposition_technology
RP+M,http://www.rpplusm.com/
Sciaky Inc.,http://www.sciaky.com/additive_manufacturing.html
SLM Solutions,http://stage.slm-solutions.com/index.php?index_en
Stratasys Direct Manufacturing,https://www.stratasysdirect.com/
Trumpf, deposition line, product,http://www.us.trumpf.com/en/products/laser-technology/solutions/applications/surface-treatment/deposition-welding.html
Voxeljet,http://www.voxeljet.de/en/

Metal Powder Manufacturers

Additive Metal Alloys,http://additivemetalalloy.com/
AP&C, Advanced Powders and Coatings,www.advancedpowders.com
Carpenter Technology Corporation,http://cartech.com/
H.C. Starck,https://www.hcstarck.com
Hoeganaes Corporation,www.hoeganaes.com
LPW Technology,http://www.lpwtechnology.com/
Materials Technology Innovations Co., Ltd.,www.mt-innov.com
Metalysis,http://www.metalysis.com/
Nanosteel Company,https://nanosteelco.com/
Norsk Titanium,http://www.norsktitanium.no/
Praxair Surface Technologies,http://www.praxairsurfacetechnologies.com/
Puris,http://www.purisllc.com/
Sandvik Materials Technology,http://smt.sandvik.com/en/

Other useful Web Sites with Information, Articles and Links to AM

3D printer and 3D printing news,www.3ders.org
3Dprint.com,http://3dprint.com
3D printing and Additive Manufacturing magazine,http://www.liebertpub.com/overview/3d-printing-and-additive-manufacturing/621/
Additive Manufacturing magazine,http://www.additivemanufacturing.media/
Authentise,http://authentise.com/
Autodesk Within,http://www.autodesk.com/products/within/overview
Delcam,http://www.delcam.com/news/press_article.asp?releaseId=2058

Engineering.com,www.engineering.com
Insidemetaladditivemanufacturing.comhttp://www.
insidemetaladditivemanufacturing.com
Metal Additive Manufacturing magazine,http://www.metal-am.com/
Metalysis,http://www.metalysis.com/
Modern Machine Shop,http://www.mmsonline.com
TCT Magazine 3D Printing, Additive Manufacturing,http://www.tctmagazine.com/
OpenSCAD,http://www.openscad.org/downloads.html
Wohlers Associates,http://www.wohlersassociates.com/

Appendix A
Safety in Configuring a 3D Metal Printing Shop

Additive manufacturing service providers or those planning to establish their own AM capability should ensure all AM manufacturer recommendation and procedures are followed to attain and maintain the skills and controls required to assure a safe AM processing environment as the environment, safety, and health management and controls associated with these materials and processes are significant.

For many makers the best way to realize their dreams in metal is to choose a 3D metal printing service provider based upon your new found knowledge provided in this book. When taking this approach the greatest safety concerns you may have would be those of the ergonomic setup of your computing space.

Small and mid-size business owners considering the purchase of a professional system, but unfamiliar with the specific hazards associated with lasers, electron beam generators, metal and powder handing should continue reading to get a sense of the additional hazards presented by the technology. The safety envelope of your shop operations may be significantly expanded beyond your current responsibilities when adding AM to your capability.

For the rest of us 3D hardware hackers and do-it yourselfers, the fun and lure of building and operating your own arc welding based system will take priority, but will need to be done safely. These folks are the same type (your grandparents?) as those building their own Heathkit stereo receiver, an Altair computer to learn BASIC programming or who took that old VW bug and turned it into a dune buggy.

Safety in Welding and Arc Based additive manufacturing systems

The best way to solve a problem is not to have one in the first place. A little knowledge can go a long way and nothing can bring your shop space to its knees or burn it to the ground faster than an unfortunate accident. Welding based AM systems present a number of opportunities to ruin your day that are not normally found in the home shop or art studio. Metal working in itself has a full laundry list of hazards and welding does a good job of rolling them all into one and adding a few more. Industrial settings provide federal OSHA (Occupational Safety and Health Association) guidelines, safety engineers and regular inspections. The maker must take the responsibility on their own to assure a safe shop and safe practice for everyone's sake. A comprehensive listing of weld safety information and fact sheets

© Springer International Publishing AG 2017
J.O. Milewski, *Additive Manufacturing of Metals*, Springer Series in Materials Science 258, DOI 10.1007/978-3-319-58205-4

can be found at the AWS Safety and Health Fact Sheets[1] and the free download of the ANSI Z49.1 Standard, Safety in Welding, Cutting and Allied Processes.[2] A full featured AM weld fabrication shop can be costly to establish and maintain. One of the best ways to avoid all this cost is too use the space provided in a properly configured fabrication environment. If you are going to go it on your own, taking a welding course is a great way to learn both the technology and safe practice at the same time. Trade schools, community colleges and maker spaces, such as TechShop,[3] are known to provide safe, well configured, work spaces staffed by knowledgeable personnel.

Knuckle draggers (like me) do not want to hear this, but welding in a residential garage attached to your house is a no-no. Fire requires three conditions, heat, fuel and oxygen. A residential garage has all of these crammed into a small space where flying sparks and combustibles make for a dangerous combination. A large well ventilated shop space, with a cement floor, free from combustibles is a good start to reduce these risks. Arc generating and spark producing operations, to include grinders or cutting torches for finishing operations, can throw sparks 30 feet or more. An automated AM or 3D metal printing arc based system may offer the potential for unattended operation, but may be hazardous without the proper engineering controls to implement a shutdown if a problem occurs. Even a brief distraction during an automated operation can create uncontrolled hazards. Attending the build process and having a proper fire extinguisher nearby is needed.

When you are wearing goggles or wearing an arc weld shield, you cannot be watching where the sparks are going. Trash cans, gas cans, paint, newspapers and saw dust under a work bench are a disaster waiting to happen. Fire resistant clothing, gloves, long sleeves and a shirt collar buttoned all the way up are needed even for a process such as gas tungsten arc welding that produces few sparks. Makers do not want to hear this either, but to hack together a flimsy motion system with a low budget welding system, not spending the money or time to make it safe to operate, is just asking for trouble. OK, enough for the grandpa advice. Here are some information and Web links to get you headed in the right direction.

Equipment Safety

Controlling heat and fire risks begins with control of your energy source. Your arc torch or heat source should be in good condition and properly set up as old or damaged equipment can be a recipe for disaster. Avoid buying that equipment at the garage sale or flea market. Old arc welding equipment can have worn or broken electrical leads, connections, hoses, grounding clamps and bad contacts. Make sure

---

[1]AWS Safety and Health Fact Sheets,https://app.aws.org/technical/facts/, (accessed March 30, 2015).
[2]ANSI Z49.1 Standard, Safety in Welding, Cutting and Allied Processes, Accredited Standards Committee Z49, Safety in Welding and Cutting,https://app.aws.org/technical/AWS_Z49.pdf, (accessed March 30, 2015).
[3]TechShop web site,http://www.techshop.ws/, (accessed March 30, 2015).

your equipment is in good shape and set up according to manufacturer's specifications or recommended practice.

During welding, molten metal, sparks, hot molten spatter and uncontrolled flames or arcs can ignite combustibles near the operation. It is amazingly easy not to notice an unintended fire due to welding while behind a welding shield. A good torch fixture and a work area clear of combustibles is a better solution than assuming you can control sparks or spatter.

After welding, care should be taken to allow the welded part to cool to prevent new fire ignition or burns during handling. Ultraviolet light from the intense welding arc requires additional controls such as welding curtains, enclosures, and eye protection. Other spark producing operations such as grinding or cutting can be just as bad or worse. The list goes on (e.g., lasers are discussed later) so be sure to read the safety information supplied by the vendor for your specific piece of equipment and the materials you will be welding.

Automated operations, such as an arc based 3D welder that do not require operator interaction, may lead to inattention where sparks or equipment shorting or unintended eye damage can occur. Just because a 3D automated deposition may not require an operator in the control loop, does not mean a knowledgeable operator should leave the process unattended.

Control of fuel or combustibles is another key aspect of fire prevention. Control of combustibles in the work environment is very often a problem for a multi-use environment. Metal chips and cutting fluids combined with welding or other spark or heat producing operations can be just as hazardous as wood or flammable liquid. Saw dust, machining oil, metal chips, gasoline, paint, and solvents have no place anywhere near a welding operation. The best solution is perform welding in a dedicated welding area free of waste baskets, solvents and away from combustibles.

Inert and Pressurized Gases

A wide variety of industrial gases may be used in welding, each presenting their own hazards. Argon, helium, carbon dioxide (CO_2) and nitrogen may be used to shield and protect the molten metal from air and oxygen to prevent oxidation, slag formation, and other problems. They also can shield air and oxygen from your lungs, which is a bad thing. As an example, argon is heavier than air and can settle in your lungs and kill you. This is more likely to happen in a confined space, but in a small enough shop without proper ventilation it is a real hazard. These gases are generally delivered in large pressurized steel bottles. These pressures can range up to 3000 psi and can depressurize in spectacular ways as demonstrated in industry over the years. So, keep the bottles chained with security caps on when in storage, chained to a cart or stand when in use. Secure outdoor storage is recommended. Be sure to vent the pressure lines after use. Gauges, lines, hoses, and fittings should be of a type approved for the gases and application conditions. Your local welding supply store should be able to set you up with the proper hardware. Matheson TriGas, an industrial gas supplier, has a good publication on the safe handling of compressed gas that can be found at this link[4] or that provided by Harris Products Safety Guidelines.[5]

Enclosed volumes can inadvertently store large amounts of pressure when heated or create low pressure (vacuum) conditions upon cooling. This can lead to a burst or crushed container due to the relative pressure difference related to heating and cooling. Designed pressure systems, utilizing welded fabrication to a certified code, or those inspected and certified by pressure testing are typically beyond the scope of the maker or hobbyist. Homemade inert gas chambers or enclosures can hide a host of hazardous conditions. Confined spaces such as inert atmosphere chambers are commonly used in 3D metal printing equipment. For multi-use shop areas, appropriate lock out and tagging and control of operations may be needed. Knowledgeable operators with current up to date skills present in the workshop are the best insurance for safe operation and to insure the appropriate levels of hazard control.

Fumes, Particles, Dust and Vapors

In addition to the industrial gases mentioned above there can also be significant hazards associated with fumes or particles generated by the process or vapor of associated products such as cleaning chemicals. Fumes associated with arc welding or any metal melting process can consist of vaporized metal, fluxes and chemicals present in the filler metal or electrode coating. Commercially available fume extractors can collect and filter out fumes and particles in work areas. Fume hoods or ventilated enclosures such as what one would build around a 3D metal printer would be used to help prevent contamination of the work area. In an industrial setting these enclosures and hoods are inspected and tested for proper configuration and flow for a given set of welding operations by a trained occupational safety professional.

Other sources of fumes produced by AM metal processes may include vaporized and condensed metal droplets, dust, dirt, flux, and soot. A large well-ventilated work area is a good place to start. Knowing what material you are depositing and what the hazards are is important. Be sure to read the MSDS Material Safety Data Sheet and vendor information provided for each material.

It is important to choose the correct cleaning method for your equipment is you have a chamber that needs cleaning with solvents, there may be a hazardous buildup of vapors. Fume, particle and vapor filters and extractors are specifically designed to your application so consult your welding supply store to assure you have the

---

[4]Matheson TriGas Inc. publication, Safe Handling of Compressed Gases in the Laboratory and Plant, 2012,http://www.mathesongas.com/pdfs/products/guide-to-safe-handling-of-compressed-gases-publ-03.pdf, (accessed March 30, 2015).
[5]Harris Products, Common Resources, Safety Guidelines, http://www.harrisproductsgroup.com/en/Technical-Documents.aspx, (accessed June 2, 2017).

correct equipment for the job. Correct type and correctly fitting vapor or particle masks are required for some applications so be sure to choose the correct one, wear it and store it correctly. The Lincoln Electric Company has an excellent publication Frequently Asked Questions Welding Fume Control.[6]

Toxic Materials

Some materials such as zinc, chromium, cadmium plating, brass or lead can produce toxic fumes when welded. While you are not likely to be 3D printing these materials, more common metals such as stainless steel can contain alloying elements such as chromium or vanadium that can vaporize and chemically react with air to produce other toxic compounds. MSDS or material safety data sheets are available on the Web and should be consulted when handling or welding wire and materials whose hazards you are unfamiliar with. A good link to this information can be found at OSHA, Occupational Safety and Health Association.[7]

Electrical Hazards

Electric shock can kill by stopping your heart while electrical burns or loss of muscle control can lead to falls. Welding equipment should be installed professionally with correct cords, plugs, fuses and electrical service. All the major weld supply company Web sites provide electrical safety information. Here is a link to related information provided by the Lincoln Electric Company Welding Safety pages.[8] Additional information may be found in (ANSI/AWS 2005)*Safety in Welding, Cutting and Allied Processes*, ANSI Z49.1.

Mechanical Hazards

Injury can occur during lifting or moving of heavy parts so remember to wear sturdy shoes, good gloves and lift only what you easily manage while bending your knees. Rotating equipment such as used for grinding or finishing can catch, snag or entrap tools, body parts, etc. Keep the area clear of tools and tripping hazards around mechanical equipment. Automated motion system and CNC control devices can also suffer from break down or incorrect programming resulting in unplanned motion. Proper guards and shielding should be used and the proper interlock controls should be in place to assure shut down when safety conditions are violated. Commercially integrated systems generally adhere to proper national safety codes but when hackers and makers are given free rein to kluge together a system it can be too easy to cut corners due to cost and inexperience.

---

[6]Lincoln Electric Company publication, Frequently Asked Questions Welding Fume Control, (2014).http://www.lincolnelectric.com/assets/us/en/literature/mc0831.pdf, (accessed March 30, 2015).
[7]OSHA, Occupational Safety and Health Association,www.osha.gov, (accessed March 30, 2015).
[8]Lincoln Electric Company web page, Welding Safety FAQ's Electric Shock,http://www.lincolnelectric.com/en-us/education-center/welding-safety/pages/electric-shock-faqs.aspx, (accessed March 30, 2015).

Eye Hazards

Good eye protection is a must in a metal and weld shop. Proper eye protection for arc welding should be clean, well-fitting and not pitted or scratched. A proper welding shield, with lenses of a density appropriate to provide comfort for the welding conditions are required. Arc flash can cause painful burns of the eyes from even a brief flash exposure. Modern automatic darkening shields work great as long as the battery is charged and they are turned on. Safety glasses offer no protection from grinding grit if you are not wearing them. In the case of automated operations shield enclosures, barriers or welding curtains may be used to protect you or others from the arc light, even if not directly involved with the operation.

Acute versus Chronic Exposures

Fumes, noise, arc burns, dust and toxic materials can lead to acute or chronic exposures. It is important to understand the hazards of each specific to your process and materials. A sudden large encounter with toxic materials such as breathing concentrated fumes or airborne powders can present acute exposures outside the range of normal processing conditions. They may be unplanned events that can overwhelm PPE (personal protective equipment) and create immediate injury. Chronic exposures are smaller level exposures over longer periods of time that can injure in less obvious ways. Poorly maintained PPE or operator complacency can neglect the dangers of chronic exposure. Either way, the operator needs to be knowledgeable of all the acute and chronic hazards present in their specific workshop operations and protect themselves and others accordingly.

Metal Powder Hazards

Powder handling or post build finishing of a 3D metal printed part can be a dirty business. Depending on the materials and processes you are using you may be producing lots of dust, vapors, and particles, ranging from mildly hazardous to toxic. Be sure to know what materials you are working with and be sure and consult the MSDS (material safety data sheet). Proper process location and control at the source may employ ventilation fans, HEPA (high efficiency particle air) filtered vacuum cleaners, fume extraction, or particle blasting chambers. Proper control and protection of personnel will include PPE, personal respirators, dust masks, and gloves, when handling transporting or cleaning powder from a fabricated part. Proper storage, cleaning and waste handling procedures are needed.

Metal powders can get everywhere and not only create an inhalation hazard but also present an explosive hazard. In addition, solvents are often used to clean chambers and surfaces prior to and after processing. Use of protection measures are important as chronic inhalation hazards are common in the shop but poorly understood or often left uncontrolled. Finely divided metal can burn or explode requiring the use of electrically grounded floor mats to control static spark discharges.[9] Another resource is National Institute for Occupational Safety and Health (NIOSH).[10] As an example, the Aluminium Association Web page has links to video for safe handling of aluminium fines and powders.[11] There are documented examples on the Web of AM metal printing companies being cited for fires and explosions resulting in injury as a result of improper metal powder processing activities.

Laser Hazards

Laser processing includes most of the hazards detailed above such as electrical, fire, mechanical, but also presents a whole unique set of hazards associated with the high energy, invisible, direct impingement, specular and diffuse reflections of coherent beams and laser light. Anyone working directly with laser producing equipment should take a formal class in laser safety based upon the American National Standard for Safe Use of Lasers (ANSI 2000) and follow all of the manufacturer's safety guidelines. Manufacturers of AM metal systems provide Class I enclosures, that is laser light containment enclosures to assure safe operations. Maintenance and repair of these systems may require special operations in which open laser light and beams may present hazards. These hazards must be mitigated through the safety procedures provided by the manufacturers of the equipment and are best left to trained service reps.

Electron Beam Melting and Welding Equipment

EBW equipment poses a number of hazards unfamiliar to most people. For one, the impingement of the electron beam upon metals can generate x-rays, so any chamber shielding installed by the vendor must remain in place during operation. High voltage up to 175 kV may be generated in certain systems, requiring special grounding and strict maintenance procedures to ensure safety during routine service and repair. A good source for safe operations and service of EBW systems can be found in the vendor's literature and the AWS Recommended Practices for Electron Beam Welding (AWS C7 Committee on High Energy Beam Welding and Cutting 2013). Maintenance and repair of these systems will require special operations in these hazards must be mitigated through the safety procedures provided by the manufacturers of the equipment and is best left to trained service reps.

---

[9]US Dept. of Labor, OSHA Web page,https://www.osha.gov/dsg/combustibledust/guidance.html, (accessed March 30, 2015).

[10]National Institute for Occupational Safety and Health (NIOSH) Web site,http://www.cdc.gov/niosh/contact/, (accessed March 30, 2015).

[11]http://www.aluminum.org/resources/electrical-faqs-and-handbooks/safety, (accessed March 30, 2015).

Appendix B
Exercises in Metal Fusion

This book has provided a lot of information on the basic building blocks of AM systems. Reading about it is one thing but having the opportunity to get some hands on learning takes it even further. For those of you with access to some simple arc welding equipment here are some learning exercises that will help you directly experience what you have learned. These examples will demonstrate interaction of localized heat sources with metal, the effects of heating and melting, the dynamics of the molten pool and the effects of cooling, shrinkage, and distortion. You will also get a sense of how parameters such as speed and power affect your ability to control the weld pool.

First you need to gain access to some welding equipment. The best type is a light duty gas tungsten arc system. If this is your first try, be sure to have someone provide you with the operational and safety principals. Better yet, have them help you setup and perform these exercises.

You can start to buy obtaining a couple dozen 2″ square pieces of 16 gauge cold rolled mild steel and a couple dozen 2″ square pieces of 0.25″ hot rolled mild steel sheared and degreased. They do not have to be square, they do not have to be these thicknesses but they do need to be bare clean steel, degreased, without rust or a coating like paint. The size is also not important but if it is all one sheet or plate, it will rapidly become distorted and you will end up discarding the stock before it is used up.

Spot Weld on Thin Sheet

A good place to start is by melting a spot weld on a thin sheet of steel with a GTAW torch and watching it heat, melt and then cool down after extinguishing the arc. Place the piece unclamped on the welding bench, put on your PPE, get comfortable, start the torch and with a low current arc begin to heat a spot. Notice changes in the plate color, shape or texture. Watch it melt, watch the sample piece move and distort while heating. You can often see a ring of moisture or burned off surface contaminants create an expanding ring changing color or luster around the periphery of the molten pool. You can watch the molten pool form and watch the surface shine and detect movements in the liquid metal. Move the torch around a bit and you will see a depression beneath the hottest part of the arc and follow the

© Springer International Publishing AG 2017

J.O. Milewski, *Additive Manufacturing of Metals*, Springer Series
in Materials Science 258, DOI 10.1007/978-3-319-58205-4

movement. The pressure or arc force is a combination of hot gasses and vaporized metal that create a localized depression that can induce movement in the molten pool. This is an important concept. As the pool grows you may see it sag, indicating you have fully penetrated or melted the sheet. Surface tension will hold this liquid in place until the pool volume of liquid gets too heavy, braking the force of the surface tension creating a drop through of liquid. As the molten pool begins to sag, quickly move away the torch and watch the pool solidify. After shutting off the arc current, watch carefully enough with your shield off. You may see the edges of the plate buckle as the plate cools. Look for a depression in the middle of the solidified spot. Are there any cracks in this weld crater location?

As the torch heated the plate, the metal expanded, as the temperature approached melting, the metal weakens and stopped expanding until all metal strength was lost upon melting. At melting you have a molten pool surrounded by a ring of orange-hot but un-melted weak, softened metal surrounded by an expanding ring of heated metal in compression due to metal expansion. Upon removing the heat source the pool solidifies and the soft hot weak orange metal begins to contract or shrink upon cooling. The formerly expanding ring shaped region of compressive forces reverses into a ring region of contracting tensile forces until the piece has fully cooled. These dynamic stresses caused by expansion and shrinkage warp and buckle the plate during welding. Those forces, relieved by movement, result in distortion and those not relieved can be locked up as residual stress remaining within the part. This is another important concept, as these complex thermo-mechanical processes are present in all welded structures to one degree of another. The smartest material scientists and mechanical engineers cannot fully quantify these effects, either by measurement or simulation but have for the most part learned where they occur and how to control them. Makers simply need to know they exist and get a feel for which way and how much a part might deform as a result of welding (or AM) so they may better plan their part design or build schedule.

Repeat this spot melting again, this time let it melt all the way until it drops through. Repeat another spot, this time with more torch movement in circles, back and forth. Watch the puddle move. This time add a drop of weld metal filler. Watch it melt and drip into the pool, dip the end of the wire directly into the pool and let the pool melt the filler. Tilt the torch and watch the pool shape change.

Linear Weld on Thin Sheet

Now that you have captured the look and feel for a stationary heat source and spot weld you can begin to move the torch and create a linear weld bead. As before, start by melting a spot. The trick to a good weld start is not to begin to move right away, but to dwell long enough for the initial weld pool penetration to be established. Lack of fusion defects is often created at the weld start or stop region. Turn the plate over and look at the melt track from the back side. Can you see increased weld penetration as the weld begins due to heat buildup in the plate?

If you are adding filler by hand wait until the correct pool size is established then begin to move and add filler by dipping the rod into the molten pool to melt.

Holding the filler close to the arc provides an indirect preheating and makes melting the filler in the pool easier. A cold filler rod can tend to quench the pool and makes it harder to deposit a uniform bead. For weld processes that automatically feed wire such as GMAW, a smooth start to the bead is a lot tougher as weld filler begins to feed immediately. You either end up with a big lump of deposit or some lack of penetration along the start of the weld joint. Readers may recall the sections relating to robotics and shape welding.

Before you start adding filler just run the weld bead along the piece to begin to develop the feel and control of moving while maintaining a constant torch position, torch height, and travel speed. Halfway across the plate stop and watch the plate distort during cooling. Finish your weld bead and observe the final shape of the plate. On another plate start right at the edge and watch how much faster the pool forms. Weld all the way across the plate until you reach the other edge. Watch how the weld pool gets larger and the heat builds up as you approach the edge. Run some more weld beads without filler but this time make some changes to torch height, torch angle and speed. Observe how the pool forms a tear drop shape and gets longer and narrower as you move faster. Stop the weld and study the weld crater at the end of the weld.

Add some filler to the leading edge of the pool and watch it flow around and get pushed about by the heat source. Watch it draw up into a bead and watch how the weld ripples change as a function of how often you add drops of filler. The top weld surface is not perfectly flat but tends to form a curved convex top surface due to surface tension forces. All these processes take place to one degree or another in AM unless you are sintering the material.

Stop the weld and start again in the crater left by the first weld. At the termination of the weld, practice filling the crater with an extra drop of filler to get a well fused end spot. Take a wet rag and selectively quench a location and watch the hot metal shrink and distort as it cools.

Partial Butt Weld and Root Break Penetration Study

To begin, take some of the thicker plate material and two plates side by side without clamping or tacking. Watch what happens during welding to the weld joint and the alignment of the two plates. Now clamp two plates side by side and make a partial penetration weld along the straight butt weld seam. Watch the joint open up as surface tension draws the molten edges of the joint apart. Watch as the melted end of the filler rod is drawn into a ball or droplet shape also by surface tension. Observe the shape and distortion of the plate after welding. You can put the piece in a vice and with a pair of vice grips or a hammer try to bend along the face of the weld. Then bend the other direction against the root of the joint and you are likely to see the weld break and tear as it much weaker when subjected to a load from the back side. Do this again at the same weld setting but this time fixture the plate with a small gap about 1/16" wide (for a ¼" plate). Try a root break again and compare the depth of penetration with and without the joint gap. Watch the edge of the plate to either side of the joint melt at the leading edge of the weld pool. Finally flip one of these welded plates over and weld it from the other side as well. You will see the distortion produced by the first weld is partially counteracted by the distortion created by the back side weld. Use a metal brush to clean any oxidation and discoloration off the weld surface to get a better look. How smooth and consistent is your top bead? Do you see any other defects as mentioned earlier?

Clamp and tack some plates in a Tee joint configuration, weld a few plates at different torch angles and speeds. Try placing a couple 1 inch long stich welds on one side of the joint then welding the other side completely. Note the reduction of distortion by offsetting opposing welds. Stand two plates up and tack the two corners on the top outer edges, weld along the top edge. Stop and observe the distortion compared to the butt weld. Look at the rounded shape of the weld bead. Clamp a plate flat on the welding table and run a weld along the outer edge and into a corner and back out along the next edge. Did you observe the buildup of heat as you entered the corner? Did you need to compensate your speed as you entered and left the corner? Did the corner round off or melt away as your weld puddle approached?

Clad Welds and 3D buildups

Demonstrate the process of building up a shape by choosing a thicker metal plate and create a clad weld by running one weld bead next to another, then another and another. Make sure to fuse each pass into the adjacent pass while penetrating into the deposit below. Then start another layer on top of that. A couple things you will notice right away are heat buildup and severe distortion. Let the deposit cool then add some more. You will find it increasingly difficult to keep the edges from rounding off. You may see some porosity and slag as oxidation of subsequent layers may begin to accumulate and form other defects. Do the same thing again but crisscross each subsequent layer. Weld in a spiral path starting at the plate center and moving outward. Weld another spiral starting around the outer edges of the plate. Make a note regarding how long it took the plate to cool off between layers and compare how long it took to deposit a layer.

Next find a large thick plate and weld a 3D pattern by tracing and building-up overlapping weld beads to form a large cylinder or 5 point star. Continue to trace the star one bead on top of another, adjusting your travel speed, arc length, and other conditions to allow 3D buildup. Wait and allow to cool, deposit some more. How high a buildup were you able to deposit?

Makers can benefit from these exercises as now you know some of what AM processes are up against. All these effects happen to one degree or another when using an AM process, only on a smaller small.

So that is it. Go out there, get some representative material and start making welds. As questions arise you can dig deeper into the details of your materials and processes to expand and refine your techniques. With a working understanding of the process and knowledge of the terms and underlying concepts of metal fusion you can make more effective use of zooming in on topics on the Web or having a technical conversation with the members of your local AM community.

Appendix C
OpenSCAD Programming Example

The OpenSCAD web site provides access to the CSG scripting software that may be used to create solid models and conversion to STL file format. Links to training manuals, videos, and code examples are provided. STL files may then be imported into other STL file processing software such as Materialise MiniMagics for fixing, slicing and conversion into a variety of common 3D printer formats as described in Appendix F. Here are two examples of the recreation of an obsolete tool design and the recreation of a complex design evolved in nature.

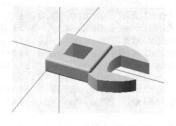

© Springer International Publishing AG 2017
J.O. Milewski, *Additive Manufacturing of Metals*, Springer Series in Materials Science 258, DOI 10.1007/978-3-319-58205-4

```
// 1/4" crowfoot Whitworth wrench
// test part for 3D metal printing

drive = 9.6; // 9.6 mm , for 3/8" drive
dpx = 10; // drive pad x
dpy = 10; // drive pad y
box_h = 6; // Height
round_r = 2; // Radius of round
dps = 65; // Number of facets of rounding sphere
sps = 200; // Number of facets smoothing spanner
spr = 1.2; // radius sphere for edge rounding
dp_offset_x = 9; // x offset of drive pad for 3/8" drive
dp_offset_z = 0.7; // z offset of drive pad for 3/8" drive
drsz = 9.6; // 9.6 mm for 3/8" drive size hole
// drsz = 13; // 13 mm for 1/2" drive size hole


// make drive pad with 9.6 mm (3/8") drive and round edge

translate([dp_offset_x,0,dp_offset_z])
{
      difference()
      {
            hull()
            {
                  translate([-dpx,-dpy,2.5]){sphere(spr,$fn=dps);}
                  translate([-dpx,-dpy,-2.5]){sphere(spr,$fn=dps);}
                  translate([dpx,-dpy,2.5]){sphere(spr,$fn=dps);}
                  translate([dpx,-dpy,-2.5]){sphere(spr,$fn=dps);}
                  translate([-dpx,dpy,2.5]){sphere(spr,$fn=dps);}
                  translate([-dpx,dpy,-2.5]){sphere(spr,$fn=dps);}
                  translate([dpx,dpy,2.5]){sphere(spr,$fn=dps);}
                  translate([dpx,dpy,-2.5]){sphere(spr,$fn=dps);}
            };
            cube(size=[drive,drive,10.0],center=true);
      };
};
```

```
difference()
{
     intersection()
     {
          translate([29,0,0])
          {
               cylinder(6,15,15,center=true,$fn=sps);
          };

          translate([17,0,0])
          {
               scale([     1.7,1,1])
               {
                    cylinder (6,15,15,center=true,$fn=sps);
               };
          };
     };

     intersection()
     {
          translate([38,0,0])
          {
               cylinder(7,14,14,center=true,$fn=sps);
          };

          translate([35,0,0])
          {
          cube([25,13.6,7],center=true);
          };
     };
};
```

```
// Rev A, Cholla Cactus skeleton.  This example models a naturally evolved
// organic structure allowing the plant to expand significantly to store
// water while providing tall rigid structure.

module lig() // build a ligament
        {
                linear_extrude(height = 34, center = true, convexity = 10,
twist = 120, $fn = 20)  //define rotation and height of ligment, enter on
X,Y
                translate([4, 0, 0])    // translate 4
                circle(r = 1.5);        // ligament diameter 1.5
        };
module leg() // build a ligament in minus rotation
        {
                linear_extrude(height = 34, center = true, convexity = 10,
twist = -120, $fn = 20)  //define rotation and height of ligment, enter on
X,Y
                translate([4, 0, 0])    // translate 4
                circle(r = 1.5);        // ligament diameter 1.5
        };

// build 4 ligs in positive rotation
for (i = [1:1:4])
{
        rotate(a=[0,0,90*i])lig();
}

// build 4 legs in positive rotation
for (i = [1:1:4])
{
        rotate(a=[0,0,90*i])leg();
}
```

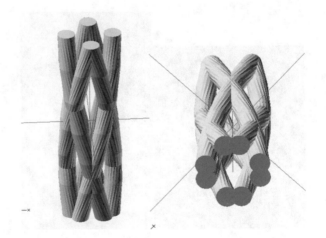

Appendix D
3D Printer Control Code Example

```
; generated by Slic3r 1.1.7 on 2014-08-04 at 14:26:25
; perimeters extrusion width = 0.50mm
; infill extrusion width = 0.52mm
; solid infill extrusion width = 0.52mm
; top infill extrusion width = 0.52mm

G21; set units to millimeters
M107
M104 S205; set temperature
G28; home all axes
G1 Z5 F5000; lift nozzle

M109 S205; wait for temperature to be reached
G90; use absolute coordinates
G92 E0
M82; use absolute distances for extrusion
G1 F1800.000 E-1.00000
G92 E0
G1 Z0.500 F7800.000
G1 X75.290 Y84.199 F7800.000
G1 E1.00000 F1800.000
G1 X75.911 Y83.722 E1.04948 F1080.000
G1 X76.499 Y83.355 E1.09322
G1 X77.152 Y83.028 E1.13929
G1 X77.890 Y82.747 E1.18921
G1 X78.556 Y82.564 E1.23277
G1 X79.154 Y82.454 E1.27119
G1 X104.832 Y78.850 E2.90857
G1 X107.018 Y78.893 E3.04662
G1 X108.368 Y79.158 E3.13351 F1080.000
```

© Springer International Publishing AG 2017
J.O. Milewski, *Additive Manufacturing of Metals*, Springer Series
in Materials Science 258, DOI 10.1007/978-3-319-58205-4

Appendix E
Building an Arc Based 3D Shape Welding System

The Michigan Tech Open Sustainability Technology Lab[12] has created a weld based 3D metal printing shape welding system for less than $2000. It uses a gas metal arc welding GTAW torch and RepRap type motion to enable 3D printed metal shapes. Design details are presented on this Appropedia.org Open-source 3D metal printer[13] Web page in this IEEE technical report[14] (Anzalone et al. n.d.).

This paper provides the information required to build your own GMAW, Rep Rap based printer but may also be used as a starting platform for other arc based type shape welding systems. A more recent report details a substrate release mechanism (Haselhuhn et al. n.d.). We'll keep our eyes on this project as it is ongoing as MTU has continued involvement in the America Makes program, (accessed March 30, 2015).

MTU correctly highlights the shortcomings of commercial AM, as high cost and low volume deposition restricts access to the technology for small to mid-enterprises and the developing world, to have an in-house capability. They provide the bill of materials and all the directions needed to build one of these machines. Near-net-shaped parts could be significantly less expensive to print, especially if the design already exists in a free database. Open-source designs and free modeling software will provide additional access to fab shops with low capitalization. Existing sources for a wide range of weld wire will reduce some of the costs and concerns regarding powder based processes. These guys are up front and realistic regarding the limitations of this demonstration level system. They do a good job providing an overview of metallurgical concerns and the need for further development of a more robust configuration. With America Makes funding and a motivated team of developers this technology is bound to make great strides in bringing model based weld deposition to AM.

---

[12]Michigan Tech Open Sustainability Technology Lab,http://www.mse.mtu.edu/~pearce/Index. html, (accessed March 30, 2015).
[13]Appropedia.org Open-source 3D metal printer, Michigan Tech,http://www.appropedia.org/ Open-source_metal_3-D_printer, (accessed March 30, 2015).
[14]A low-cost open-source 3D metal printer,https://www.academia.edu/5327317/A_Low-Cost_ Open-Source_Metal_3-D_Printer, (accessed March 30, 2015).

© Springer International Publishing AG 2017
J.O. Milewski, *Additive Manufacturing of Metals*, Springer Series
in Materials Science 258, DOI 10.1007/978-3-319-58205-4

Appendix F
Exercises in 3D Printing

A good way to get hands on experience with 3D printing is to find and download 2D and 3D CAD software and learn to create your own models, translate them into STL files and send them to a 3D print service provider. Another way is to explore the wide range of available sites that provide free access to 3D computer models and low cost 3D printing services. A third way is to download or create your own 3D model and print it on a personal 3D printer at home or at a library or other maker space available to you. These exercises may be performed using free or low cost software and plastic or other materials other than metals to reduce the cost. We'll mention links to some entry open-source level software that can be used for free.

Exercise 1 Locate and evaluate 2D and 3D modeling software.

As with the metal fusion exercises presented in Appendix B, these exercises are meant to provide the reader with the opportunity to explore 3D design creation, learn and exercise the process of locating and submitting a design to a personal printer or printing resource. In these exercises the difference between 3D printing in plastic and metal will simply be a menu selection and a cost difference.

A listing of 2D and 3D CAD software listing may be found at the RepRap.org Web page.http://reprap.org/wiki/Useful_Software_Packages#2D_and_3D_CAD_ software,http://reprap.org/wiki/Useful_Software_Packages

Flat scan to 2D CAD vector to DXF to OpenSCAD, can be found at this link. http://3dprint.com/54483/reverse-3d-printing, this allows more complex geometry to be designed in 2D Cad and imported into a parametric 3D CSG scripting environment.

Search the Web for other available packages using search terms such as "3D CAD software", "3D solid modeling software", "open source 3D CAD", "open source 3D modeling software" or search YouTube for videos and tutorials.

Exercise 2 Locate and evaluate STL software for model checking, fixing and slicing.

One such example is the free STL viewer, MiniMagics from Materialise,http:// software.materialise.com/minimagics, (accessed April 25, 2015).

One example of STL slicing software is Slic3r,http://slic3r.org/, (accessed April 25, 2015).

© Springer International Publishing AG 2017
J.O. Milewski, *Additive Manufacturing of Metals*, Springer Series
in Materials Science 258, DOI 10.1007/978-3-319-58205-4

A listing of STL software both open and closed source may be found at the RepRap.org Web pagehttp://reprap.org/wiki/Useful_Software_Packages

Search the Web for other available packages using search terms such as "STL software", "open source STL slicing software", "open source STL", or search YouTube for videos and tutorials.

Exercise 3 Access a 3D printing service provider, locate a model and get a quote.

On your PC go to the Web and search for "3D printing service" to locate a number of sites that can provide a service. Explore some of these sites to see the range of offerings. Some of these sites offer viewing and shopping for model designs that may be bought, printed and shipped to your home or business. Some offer student discounts while many offer instructional videos to get you started. Explore the wide range of materials and colors available. Some offer metal printing as well. They allow you to set up an account, submit STL models and get a quote. To submit a model this usually involves creating a free account with name and password. They often offer translation from many different CAD formats into STL format. STL file checking and fixing are offered as well as design guides and expert advice regarding worthiness of the design for printing with respect to material, size, and design features. Samples of materials are offered from some sites (plastics and polymers) with some sites offering small example parts. Some sites offer additional design services and links to user groups, social groups, and community forums. Links to software and free ware may be used to download a design, modify it and learn the software. Customization of existing designs and products is also provided. Links to professional design services and programming services such as API (Application program Interfaces) are offered. Online stores allow you to become a seller of your designs or create a virtual online factory.

As you will see many of these sites specialize in plastics, polymers, and materials other than metals. When you submit model designs for quotes you will see a fairly wide range of prices associated with material types and the desired finishing options. Be aware that if you order the "same" part from different vendors they may not look exactly the same. You will also notice size limitations and delivery time variations amongst vendors. Plastics are a good place to start due to cost but recall from what you learned in the book. There can be significant difference between these processes and especially between metal and non-metal processes. Plastics and open-source software is a good place to begin to learn if you are a novice, but do not get lost in the technology as there is an opportunity cost associated with learning a lot about a technology you ultimately won't be using.

https://www.stratasysdirect.com/promos/3d-printing-services.html?gclid=CIeigJn
418QCFQcvaQodJmIA7g
http://www.shapeways.com/
http://www.ponoko.com/3d-printing,
http://i.materialise.com/,
http://www.sculpteo.com/en/services/

Exercise 4 Locate AM metal printing service providers and evaluate the options.

Service providers specializing in AM and metals take these service offerings to a higher professional level. Costs for printing and materials increase significantly as well. You won't find the social links or online stores or market places available and most products are not consumer products unless you are shopping for jewelry or small demonstration items. As described earlier in the book, materials are limited and a wider range of post process finishing options are available such as heat treatments and HIP processing. Design guides are available online and other professional services are provided or available as references. Part quality is also taken to higher levels with certified materials and in some cases adherence to industry standards. Some sites offer instant quotes but most offer professional consolation in conjunction with uploading a candidate design model. Many of the service providers currently available also provide conventional prototyping services using casting and CNC machining. Web searches should be used to locate new additive manufacturing and rapid prototyping services offering a greater selection to choose from. Case studies and example parts are often featured on these Web sites. They may also provide links to material data sheets. A selection of AM machine vendors is listed in the AM Machine Builders portion of the book. Often the machine vendor's provide web links to preferred service providers. Another good source of current service providers is maintained on the Wohlers Associates Web site.

https://www.solidconcepts.com/3d-printing/,
http://www.protolabs.com/direct-metal-laser-sintering/
http://www.nextlinemfg.com/metal-3d-printing/?gclid=CNK_yd2H2MQCFYVAa
QodHaIA6w
http://www.protocam.com/metal-prototyping/,
https://www.stratasysdirect.com/materials/direct-metal-laser-sintering/

Appendix G
Score Chart of AM Skills

Given your needs AM interest and needs, how would you weight this listing of your skill set? Are there successful examples of companies or individuals within your industry or field that may be used as examples to pin the weights and rankings? Non-applicable (NA) selections when chosen are not figured into the ranking. As an example if you will always be using the CAD designs provided by others you would select NA when assessing your existing CAD capability of need to outsource CAD design function.

Design Engineering skills (high, mid-level, low, NA, low outsource, high outsource), (6, 3, 1, 0, −3, −6)

Relevant inspection capability (Full, some, neutral, low outsource high outsource), (4, 2, NA, −2, −4)

CAD, (full, some, none, outsource some, outsource all, NA), (6, 3, 0, −3, −6, NA)

CAM, CNC tools, (full, some, NA, outsource some, outsource all), (6, 3, NA, −3, −6)

3DP, STL, RP experience with plastics, other materials (full, some, none), (4, 2, 0)

Conventional metal fabrication, (full, some, none), (4, 2, 0)

Relevant materials or metallurgy experience (high performance materials, powders, weld alloys), (2, 2, 2)

Relevant processing experience, (forming, casting, machining, welding, CNC, PM), (1, 1, 1, 1, 1)

Relevant in house post processing, (machining and cutting, EDM, surface finishing, HT, HIP), (3, 4, 2, 4, 5)

Relevant industry or market sectors (aero, marine, auto, medical, energy, tool-die, prototyping), (1, 1, 1, 1, 1, 1, 1)

Existing customer needs (prototyping, R&D, small lot production, existing product lines, recurrent), (2, 2, 2, 2, 2)

What specs, codes or regulations and levels of formality do you work to? (Strict, moderate, low, none), (3, 2, 1, 0)

What floor space and facility configuration is available? (Existing, configurable, expansion, none), (3, 2, 1, 0)

© Springer International Publishing AG 2017
J.O. Milewski, *Additive Manufacturing of Metals*, Springer Series in Materials Science 258, DOI 10.1007/978-3-319-58205-4

Glossary

An important goal of this book is to familiarize readers with terms and jargon found in the articles, books, and papers associated with AM metal. This jargon has changed and evolved therefore many of the terms are increasing in usage while others are decreasing. This listing is not meant to compete with the standardizing of terms and definitions for AM by national and international organizations (ASTM 2012), or ISO/ASTM 52900, Additive Manufacturing— General Principals and Terminology, as those are bound to evolve as well. .

Abrasive wear the in-service erosion of a surface due to friction and contact.

Additive manufacturing a general term describing a manufacturing process that adds material rather than removes material. The common usage currently relates to near-net-shaped 3D parts beginning with a 3D computer model and the joining of feedstock, often using powder, wire or liquid, layer by layer, to form a 3D part.

Aging a metallurgical process where the properties of some alloys can change over time at near ambient temperatures, and is a much slower process in comparison to heat treatment.

Artificial aging an acceleration of the aging process by the introduction of heat.

Alloy metallic substance of two or more elements, at least one of which is a metal.

Amorphous metal a non-crystalline metal structure often formed by rapid cooling.

Anisotropy, Anisotropic a change or variation of structure and properties as a function of direction within a bulk material, dissimilar or non-uniform in all directions, as in anisotropic properties, also see*isotropic*.

Annealing a heat treatment allowing crystal structures to reform, relax and relieve internal stresses.

As-built parts AM parts, with base plate or support structure removed, without other post processing steps.

© Springer International Publishing AG 2017
J.O. Milewski, *Additive Manufacturing of Metals*, Springer Series
in Materials Science 258, DOI 10.1007/978-3-319-58205-4

Atomization the creation of spherical powder particles using melting and solidification.

Balling an undesirable, irregular surface, deposition track, or deposit bead feature.

Base component an existing part upon which AM features are deposited, such as in repair.

Base feature an existing featured component upon which AM features are deposited.

Base plate a metal plate within a build chamber upon which AM support structures or parts are deposited.

Bespoke custom made, made to order such as clothing or personal items.

Blade crash in the PBF processes, the collision of a powder recoating blade with the structure being built.

Bounding box the geometric extent of a 3D model oriented within a build volume.

Build chamber a sealed, often inert or vacuum enclosure within which the part is built.

Build cycle the entire build sequence from preheat to cool down.

Build job CAM file ready to be loaded and executed.

Build orientation the geometric X, Y, Z relationship of the part to the build volume reference frames.

Build platform or base plate, upon which a part and support structures are deposited.

Build-rate usually in cubic centimeters per hour, describing the average time to build an AM part.

Build speed see build rate.

Build sequence the sequence of machine functions needed to build the part, from preheat to cool down.

Build surface the surface upon which another AM layer may be added.

Build volume in PBF, the volume or dimensions of the powder bed.

Bulk material a volume of material displaying the characteristic of the structure and properties of the deposit.

Burned out the powder metallurgy process of removing binder from a green part using a furnace treatment.

Buy-to-Fly in aerospace, the ratio of material purchased to the final amount used in the aerospace component.

Cartesian space a volume described by X, Y, Z coordinates relative to a defined origin.

Certification Third-party attestation or issued certificate related to the quality, conformance or performance of products, processes or persons.

Cladding a fused layer of material added to a base component to impart additional surface properties.

Coincident facets a defect in an STL model, triangular surface facets occupying the same space.

Cold bond bonding without full melting and solidification.

Cold lap solidification of adjacent regions without full mixing or breakup of oxide layers during solidification.

Carrier gas or delivery gas, often inert or non-reactive to deliver powder feedstock using a pressurized feed.

Casting an additive metal fabrication process that relies on introduction of molten material into a mold.

Coalescence the joining of melted or liquid regions to form one, also the joining of gas bubbles in a liquid.

Corrosion a chemical process, often undesirable, acting on a material surface, changing its properties.

Consumable electrode a current carrying wire form of filler, melting and transferring material across the arc.

Contouring in AM PBF, the following of the perimeter or outline of the part within a 2D slice.

Contour path see contouring, often one of multiple paths, offsets or other conditions may be defined.

Cracking (hot, cold, solidification, crater), a localized tearing or separation featuring initiation and propagation.

Crater a depression region at the end of a melt pass with localized shrinkage stresses and sometimes cracking.

Crowd sourcing in AM, an informal self-assembly of talent gathering to jointly create or solve a technical issue.

Crystal an ordered structure of atoms or molecules that may self-assemble or grow displaying long range order.

Curling in AM shrinkage in the XY plane that creates distortion in Z, often a peeling off of the build platform.

Defect an undesirable, discontinuous region violating the acceptance criteria.

Densification the closing of voids and porosity within a deposit.

Deposition path or track, the path of laser, electron beam or other heat source used to sinter or fuse the material.

Deposition rate the rate at which feedstock is fused or sintered into the part, often defined in cc/h or lbs. /h.

Deposition sequence see build sequence.

Depth of fusion the depth to which the molten pool has penetrated into the substrate.

Diffusion movement of an atom or molecule from a region of high concentration to low concentration.

Discontinuity a discontinuous region of the deposit not violating the acceptance criteria.

Down facing surface see overhang.

Ductility the ability of a material to be deformed when pulled and stretched.

Dwell time a delay in the motion sequence, allowing other sequence function, such as preheat, to occur.

Elastic limit the point at which a stretched material will be permanently deformed and not spring back to shape.

Electron beam a directed stream of electrons produced by an electron generating source within an electron gun.

Elongation the length change in percent of a material due to tensile loading and stretching to failure.

Energy density the amount of energy per unit area, often expressed in W/mm^2.

Epitaxial growth grain growth during solidification of a preferred orientation based upon neighboring grains.

F number orF# a term used to describe the convergence condition of a focused laser beam.

Fabricate to produce a part from the processing of raw materials.

Facet in AM and STL file format, a single triangular approximation of CAD model surface.

Feedstock often powder or wire in AM, the raw material form processed to form the object.

Flaw see discontinuity.

Fully dense without significant content of voids.

Functionally graded a change in the AM deposited material chemistry, structure and properties of deposited material resulting in the desired function of a part feature and its location.

Grain a bounded region of a similar crystallographic phase, chemistry and orientation.

Grain structure the characteristic phase, chemistry and defect morphology within a grain and bulk material.

Green part a part made up of partially sintered powder.

Green shape see green part.

Hard facing see cladding, cladding with a hard material to accommodate wear or impact.

Hardness resistance to indentation, wear or impact, often related to strength.

Hatch lines adjacent, offset deposition paths, within a planar or surface layer.

Hatch pattern the orientation of adjacent hatch lines or contour paths within a layer or between layers.

Hatch spacing the offset between hatch lines.

Heat treatment heating a part to a temperature below melting, long enough to induce microstructural changes.

Homogenization a heat treatment used to allow an equalization of microstructure and chemistry.

Hog out in machining, the removal of a large amount of material to form a cavity, resulting in material waste.

Hot Isostatic pressing a process applying a high pressures and temperatures below melting to consolidate powders and close voids and porosity.

Humping a welding term related to an irregular top surface of a weld bead.

In-situ in place, in process.

Infiltration a thermal process used to fill voids within a porous structure using a lower melting point metal.

Intensity profile the variation of laser or electron beam intensity within a focal spot of impinging beam.

Interaction zone in AM, the localized region of energy above and below the focal location capable of melting.

Interstitial elements elements within a crystal or grain that are not part of the ordered structure.

Inverted normal in STL file format, the orientation of a triangular facet pointing *inward* to the solid shape.

Isotropic similar or uniform in all directions, as in isotropic properties or behavior.

Keyhole a vapor cavity formed in a molten pool of metal by a high energy beam.

Lack of fusion failure to coalesce a melted interface prior to solidification leaving a void.

Laser an optical device to transform light into a coherent high energy beam of photons.

Laser cladding cladding using a laser heat source.

Laser glazing laser surface melting.

Laser intensity profile spatial energy density at a location within a laser beam path.

Layer adjacent tracks of deposited material within a single slice of a model, fused together across a planar or part surface.

Layer thickness the depth of Z-motion, offset from previous layer path, or depth of a single powder recoat layer.

Liquation the partial remelting of alloying constituents in a previously deposited layer or substrate.

Manufacturing to make or process a finished component often relying on the assembly of fabricated parts.

Metallic Bond the weak chemical bonds of metallic elements featuring electron mobility and metallic properties.

Meta-stable phases complex crystallographic phases that can change over time, altering bulk properties.

Microstructure the crystal structure, grain structure and defect morphology characteristic of the bulk or a region.

Multi-physics the combination of first principal models such as mechanical, thermal or fluid flow to model AM.

Multi-scale the combination of models of the atomic, microstructural and macrostructure (part) size scales.

Nd:YAG neodymium-doped yttrium aluminium garnet, laser.

Near-net shape processing refers to the method of forming a part optimizing the use of the feed stock and minimizing the use of material, waste and post processing needed for the final part.

Non-manifold edges a defect in an STL model, where triangular facet edges are shared by more than two facets.

Open architecture in AM, the sharing, definition, specification or design of a system for use by others.

Open source publicly shared information, for users or developers.

Overhang a deposition path or region not directly above or within the boundaries of a previous layer.

Overlap when the deposition track width exceeds the hatch spacing or offset, usually defined in %.

Parametric solid model a solid geometry model where dimensions are defined as variables.

Perimeter path see contouring path.

Phase (crystallographic) material with a distinct crystal structure and uniform physical properties.

Phase (of matter) see **state of matter**, as in solid, liquid, gas, plasma.

Porosity in AM metal, a void, often spherical, formed by gas evolved from the melt during solidification or the remelting of a lack of fusion void.

Powder blend more than one powder lot, using virgin or recycled materials, mixed together.

Powder feed rate delivery rate of a powder feedstock, may be expressed in weight or volume / sec units.

Powder lot powder vendor supplied material of the same production batch, chemistry and morphology.

Powder necking spherical powder particles fused together during powder production or AM processing.

Powder satellites see powder necking.

Powder virgin unused powder, as received from the vendor, properly handled and stored.

Precipitation hardening a heat treatment used to grow strengthening phases within the microstructure.

Qualification assurance a person or process is able to operate and perform to a specified standard.

Quenching a rapid cooling used to lock in desirable microstructural crystalline phases.

Real time a secondary or additional process, such as monitoring or control, coincident in time with the process.

Recoat layer a thin layer of powder feedstock.

Recoating blade in PBF, a precision mechanical spreading device.

Recycle frequency the number of times virgin powder feedstock has been used in a powder bed fusion operation.

Recycle powder in AM, reused powder that has been sieved and baked out to be used again for AM processing.

Recrystallization a heat treatment used to create a uniform microstructure.

RepRap replicating rapid prototype machine, (of RepRap Project origin) open design motion and control software.

Residual stress in AM, mechanical forces locked up within a parts structure as a result of expansion and shrinkage.

Reuse powder see recycle powder.

Scanning the rapid back and forth or relative motion of the energy beam with respect to the part surface.

Scan pattern also see hatch pattern, the planned deposition paths associated with one or a series of layers.

Scan speed the relative traversal speed of the beam focal position with respect to the part surface.

Scan strategy also see scan pattern, the design and selection of scan paths to optimize conditions such as accuracy, density and control conditions such as warping curling and residual stresses.

Scan tracks the material deposited, sintered or fused along and scan path.

Secondary powder recycled powder.

Segregation the separation and localization of alloy constituents or impurities during solidification.

Shrink wrapping a software method used to make an STL file "water tight".

Single slice in STL models, a planar cross section defining the area and boundaries at a specific part location.

Slicing the process of producing a series of evenly spaced planar slices and cross sections from an STL model.

Slice thickness the offset of slices in relation to the Z axis of the part.

Smoke AM jargon describing a cloud of electrostatically charged powder particles suspended above the beam impingement region in EBM.

Soak time in heat treatment, the duration in time required for the entire part to reach a uniform temperature.

Solutionizing a heat treatment used to uniformly distribute segregated alloy constituents.

Solid State transformation crystallographic phase changes that occur over specific temperature and time ranges.

Spot size a diameter or dimension related to a focused high energy beam, *various technical definitions exist.*

Stair stepping a surface condition related to the layer height most visible on low angle surfaces.

State of matter solid liquid gas, plasma, often referred to as phase of material.

Stress relieve a heat treatment to relax locked up forces within a microstructure due to solidification shrinkage.

Substrate see build plate, or base component, may also refer to a previously deposited layer.

Support structure structure added to the part design to anchor and support the part during the build process.

Subtractive manufacturing is a general term referring to processes where material is removed rather than added, such as when machining a part.

Surface treatment a processing step used to impart specific chemical or metallurgical properties to part surface.

Teeth a support structure design to accommodate positive location during a build, help prevent curling or warp age and facilitate removal from the base plate and as-built part.

Topology optimization a computer based solid model FEA method, using iterative simulations, to assist the designer to remove unneeded material and reduce the weight and optimize other functional aspect of the design.

Ultimate tensile strength the strength of a material being pulled just prior to failure.

Validation objective evaluation, often by a third party, to assure the design or part meets the functional requirements and the application for which it was designed.

Verification the confirmation through the provision of objective evidence, such as with data and analysis, that the specified requirements have been fulfilled and that the part meets specifications.

Virgin powder unused powder from one powder lot, properly stored and handled.

Voids unfilled, unfused, localized defective regions within the bulk deposit, often resulting from lack or fusion or loss of material during processing, porosity is a type of void.

Warping curling or mechanical distortion resulting from thermal expansion and cooling.

Work hardening increasing the hardness and often strength of a material by deformation.

Yb:YAG yttrium aluminium garnet, laser.

Yield strength the strength of a material when pulled to it elastic limit.

Young's modulus a measure of elastic performance or the force needed to stretch a material.

References

ANSI. 2000.*American National Standard for Safe Use of Lasers*, vol. Z136.1. American National Standards Institute.

ANSI/AWS. 2005.*Safety in Welding Cutting and Allied Processes*, vol. Z49.1. Miami, FL: American Welding Society.

Anzalone, Gerald C., Chenlong Zhang, Bas Wijnen, Paul G. Sanders, and Joshua M. Pearce. n.d. A low-cost open-source 3-D metal printing.*IEEE Access*. doi:10.1109/ACCESS.2013.2293018.

Ardila, L.C., and F. Garciandia. 2014. Effect of IN718 recycled powder reuse on properties of parts manufactured by means of Selective Laser Melting.*Physics Procedia* 56: 99–107.

ASTM. 2012.*ASTM Standard Terminology for Additive Manufacturing Technologies,* Vols. F2792-12a. West Conshohocken, PA: ASTM International. doi:10.1520/F2792-12A.

Atzeni, E, and A. Salmi. 2012. Economics of additive manufacturing for end-usable metal parts. *The International Journal of Advanced Manufacturing Technology* 62: 1147–1155. doi:10.1007/s00170-011-3878-1.

AWS A3.0M/A3.0:2010. *Standard Welding Terms and Definitions*. Miami, FL: American Welding Society.

AWS C5 Committee on Arc Welding and Cutting. 1980.*Recommended Practices for Gas Tungsten Arc Welding*, Vols. C5.5:-80. Miami, FL: American Welding Society, AWS.

AWS C5 Committee on Arc Welding and Cutting. 1989.*Recommended Practices for Gas Metal Arc Welding*, Vols. C5.6:-89. Miami, FL: American Welding Society, AWS.

AWS C7 Committee on High Energy Beam Welding and Cutting. 2013.*Recommended Practices for Electron Beam Welding*, vol. A/C7.1. Miami, FL: American Welding Society.

AWS Committee on Methods of Inspection. 1980.*Welding Inspection*, 2nd ed. Miami, FL: American Welding Society.

Bostrum, Nick. 2014.*Superintelligence: Paths, Dangers, Strategies*. Oxford University Press.

Boyer, H.E., and T.L. Gall. 1985.*Metals Handbook*, Desk ed. Metals Park, OH: ASM International.

Boyer, R., G. Welsh, and E.W. Collings (eds.). 1994.*Materials Properties Handbook: Titanium Alloys, ASM International, 1994*. Materials Park, OH: ASM International.

Brackett, D., I. Ashcroft, and R. Hague. 2011. Topology optimization for additive manufacturing. In*International Free Form Symposium*, 348–362. University of Texas in Ausitin.http://utwired.engr.utexas.edu/lff/symposium/proceedingsArchive/pubs/Manuscripts/2011/2011-27-Brackett.pdf.

Brandl, E., B. Baufeld, C. Leyens, and R. Gault. 2010. Additive manufactured Ti-6Al-4V using welding wire: Comparison of laser and arc beam deposition and evaluation with respect to aerospace material specifications.*Physics Procedia* 5: 595–606.

Chang, K.-H., and C. Chen. 2013. University of Oklahoma, 3D shape engineering and design parameterization.*Computer-Aided Design and Applications* 8 (5): 681–692. doi:10.3722/cadaps.2011.681-692.

© Springer International Publishing AG 2017
J.O. Milewski, *Additive Manufacturing of Metals*, Springer Series
in Materials Science 258, DOI 10.1007/978-3-319-58205-4

Crawford, Matthew B. 2009.*Shop Class as Soulcraft: An Inquiry into the Value of Work*. New York: Penguin.

Duley, Walter W. 1999.*Laser Welding*. New York: Wiley.

Dutta, B, and Francis H. Froes. 2015. The additive manufacturing (AM) of titanium alloys. In*Titanium Powder Metallurgy*, ed. Ma Qian and Francis H. Froes, 447–468. Elsevier. doi:10. 1016/B978-0-12-800054-0.00024-1.

Easterling, K. 1983.*Introduction to the Physical Metallurgy of Welding*. Butterworth & Co Ltd.

Erik Brynjolfsson, Erik, and Andrew McAfee. 2014.*The Second Machine Age*. New York, NY: W. W. Norton and Company.

Evans, G.M., and N. Bailey. 1997.*Metallurgy of Basic Weld Metal*. Cambridge: Abington Publishing.

Ford, Martin. 2015.*Rise of the Robots*. New York: Basic Books.

Frazier, William E. 2014. Metal additive manufacturing: A review.*JMEPEG* 23 (6): 1917–1928.

Freitas Jr., Robert A. 1980. A self-replicating interstellar Probe.*Journal of the British Interplanetary Society* 33: 251–264.

Freitas Jr., Robert A., and Ralph C. Merkle. 2004.*Kinematic Self-Replicating Machines*. Georgetown, TX: Landes Bioscience.

Fulcher, B.A., D.K. Leigh, and T.J. Watt. 2014. Comparison of ALSI190MG and AL 6061 Processed Through DML. In*International Solid Freeform Symposium*, 404–419. Austin: University of Texas at Austin.http://sffsymposium.engr.utexas.edu/2014TOC.

Furrer, D.U., and S.L. Semiatin (eds.). 2009.*Fundamentals of Modeling for Metals Processing*, vol. 22A. Materials Park, OH: ASM International.

Furrer, D.U., and S.L. Semiatin (eds.). 2009b.*Metals Process Simulation, ASM Handbook*, vol. 22B. Materials Park, OH: ASM International.

Furukawa, K. 2006. New CMT arc welding process—Welding of steel to aluminium dissimilar metals and welding of super-thin aluminium sheets.*Welding International* 20: 440–445.

Gates, Bill. 1993.*Business @ The Speed of Thought: Using a Digital Nervous System*. New York: Warren Books.

Gibson, I., D.W. Rosen, and B. Stucker. 2009.*Additive Manufacturing Technologies, Rapid Prototyping to Direct Digital Manufacturing*. New York: Springer.

Giovanni Moroni, Wahyudin P. Syam*, and Stefano Petrò. 2015. Functionality-based part orientation for additive manufacturing. In*25th CIRP Design Conference*, ed. International Scientific Committee of "25th CIRP Design Conference". Elsevier B.V. Procedia CIRP 00 (2014) 000–000.www.sciencedirect.com.

Gong, Haijun. 2013.*Generation and detection of defects in metallic parts fabricated by selective laser melting and electron beam melting and their effects on mechanical properties*. Dissertation, Doctor of Philosophy, Department of Industrial Engineering, University of Louisville, Louisville, Kentucky.

Hamilton, R.F., T.A. Palmer, and B.A. Bimber. 2015. Spatial characterization of the thermal-induced phase transformation throughout as-deposited additive manufactured NiTi builds.*Scripta Materialia*. doi:10.1016/j.scriptamat.2015.01.018.

Haselhuhn, Amberlee S., Eli J. Gooding, Alexandra G. Glover, Gerald C. Anzalone, Bas Wijnen, Paul G. Sanders, and Joshua M. Pearce. n.d. Substrate release mechanisms for gas metal arc 3-D aluminium metal printing.*3D Printing and Additive Manufacturing* 1 (4): 204–220.

Herzog, Dirk, Vanessa Seyda, Eric Wycisk, and Claus Emmelmann. 2016. Additive manufacturing of metals.*Acta Materialia* 117: 371–392.

Hornick, John. 2015.*3D Printing Will Rock the World*. Amazon: CreateSpace Independent Publishing Platform.

Irving, R. R. 1981. Shape welding: A new concept in fabrication.*Iron Age*.

Kaku, Michio. 2014.*The Future of the Mind*. New York: DoubleDay.

Kapustka, N., and I.D. Harris. 2014. Exploring arc welding for additive manufacturing of titanium parts.*Welding Journal* 93 (3): 32–36.

Kapustka, Nick. 2015, April. Achieving higher productivity rates using reciprocating wire feed gas metal arc welding.*Welding Journal* 94: 70–74.

Kovacevic, R. 1999. Rapid prototyping technique based on 3D welding. In*1999 NSF Design & Manufacturing Grantess Conference*, 5–8.http://lyle.smu.edu/me/kovacevic/papers/nsf_99_grantees_1.html. Accessed 20 Mar 2015.

Kurtzman, Joel. 2014.*Unleashing the Second American Century: Four Forces for Economic Dominance*. New York: Public Affairs books, Perseus Books Group.

Kurzweil, R. 2005.*The Singularity is Near: When Humans Transcend Biology*. Penguin Group LLC.

Lachenburg, K., Stecker, S. 2011. Nontraditional applications of electron beams. In*ASM Handbook Welding Fundamentals and Processes*, ed. T. Babu, S.S. Siewert, T.A. Acoff, and V.L. Linert, 540–544. Metals Park: ASM International.

Lancaster, J.F. (ed.). 1986.*The Physics of Welding*, 2nd ed. Oxford: International Institute of Welding, Pergamon Press.

Lewis, Gary, K, and Eric Schlienger. 2000. Practical considerations and capabilities for laser assisted direct metal deposition.*Materials and Design* 21: 417–423.

Lienert, T.J., S.S. Babu, T.A. Siewert, and V.L. Acoff (eds.). 2011.*Welding Fundamentals and Processes, ASM Handbook*, vol. 6A. Materials Park, OH: ASM International.

Lipson, Hod. 2014. AMF Tutorial: The Basics (Part 1).*3D Printing* 1 (2): 85–87 (Mary Ann Liebert, Inc).

Mahmooda, K., W. Ul Haq Syedb, and A. J. Pinkerton. 2011. Innovative reconsolidation of carbon steel machining swarf by laser metal deposition.*Optics and Lasers in Engineering* 49 (2): 240–247.

Mantrala, K.M., M. Das, V.K. Balla, C.S. Rao, and V.V.S. Kesava Rao. 2015. Additive manufacturing of Co-Cr-Mo alloy: Influence of heat treatment on microstructure, tribological, and electrochemical properties. Frontiers in Mechanical Engineering 1 (2). doi:10.3389/fmech.2015.00002.

Mayer-Schönberger, Viktor, and Kenneth Cukier. 2013.*Big Data: A Revolution that will Transform How We Live, Work, and Think*. New York: Houghton Mifflin Harcourt, Eamon Dolan Book.

McAninch, M.D., and C.C. Conrardy. 1991. Shape melting—A unique near-net shape manufacturing process.*Welding Review International*.

Moylan, S., J. Slotwinski, A. Cooke, K. Jurrens, and M.A. Donmez. June 2013. Lessons Learned in Establishing the NIST Metal Additive Manufacturing Laboratory. NIST Technical Note 1801, Intelligent Systems Division Engineering Laboratory, NIST.

Murr, L.E., S.M. Gaytan, A. Ceylan, E. Martinez, J.L. Martinez, D.H. Hernandez, B.I. Machado, et al. 2010. Characterization of titanium aluminde alloy components fabricated by additive manufacturing using electron beams.*Acta Materilia* 58: 1887–1894.

Murr, Lawrence E., Sara M. Gaytan, Diana A. Ramirez, Edwin Martinez, Jennifer Hernandez, Krista N. Amato, Patrick W. Shindo, Francisco R. Medina, and Ryan B. Wicker. 2012. Metal fabrication by additive manufacturing using laser and electron beam melting technologies. *Journal of Materials Science & Technology* 28: 1–14.

Naím, Moisés. 2013.*The End of Power: From Boardrooms to Battlefields and Churches to States, Why Being in Charge isn't what it used to be*. New York: Basic Books.

Norman, Donald A. 2004.*Emotional Design*. Basic Books.

O'Brien, A., and C. Guzman (eds.). 2007.*Welding Handbook, Welding Processes, Part 2*, vol. 3, 9th ed. Miami, FL: American Welding Society.

Palmer, T., and J.O. Milewski. 2011.*Laser Deposition Processes*, vol. 6A. In*ASM Handbook Welding Fundamentals and Processes*, ed. T. Siewert, T. Babu, S. Acoff, and V. Lienert, 587–594. Metals Park: ASM International.

Pauly, S, L. Lober, R. Petters, M. Stoica, S. Scudino, U. Kuhn, and J. Eckert. 2013. Processing metallic glasses by selective laser melting.*Materials Today*. doi:10.1016/j.mattod.2013.01.018

Pisano, Gary P., and Willy C. Shih. 2012.*Producing Prosperity: Why America Needs a Manufacturing Renaissance*. Watertown, MA: Harvard Business Review Press.

Porter, D.A., and K.E. Easterling. 1981.*Phase Transformations in Metal Alloys*. Workingham, Berkshire: Van Nostrand Reinhold Ltd.

Ready, J.F., and D.F. Farson (eds.). 2001.*LIA Handbook of Laser Materials Processing*. Orlando, FL: Laser Institute of America, Magnolia Publishing Inc.

Reid Hoffman, Reid, and Ben Casnocha. 2012.*The Start-up of You*. Random House LLC.

Sames, W.J., F.A. List, S. Pannala, R.R. Dehoff, and S.S. Babu. 2015. The metallurgy and processing science of metal additive manufacturing.*International Materials Review* 1–46.

Schmidt, E., and J. Cohen. 2013.*The New Digital Age*. New York: Alfred A. Knopf.

Steen, W.M., and J. Mazumder. 2010.*Laser Material Processing*, 4th ed. London: Springer.

Thiel, Peter. 2015.*Zero to One—Notes on Startups or How to Build the Future*. Crown Business, a division of Random House.

Todorov, Evgueni, Roger Spencer, Sean Gleeson, Madhi Jamshidinia, and Shawn M. Kelly. 2014. *Nondestructive Evaluation (NDE) of Complex Metallic Additive Manufactured (AM) Structures*. Interim, Air Force Research Laboratory, AFRL-RX-WP-TR-2014-0162.

Triantaphyllou, Andrew, Claudiu L. Giusca, Gavin D. Macaulay, Felix Roerig, Matthias Hoebel, Richard K. Leach, Ben Tomita, and Katherine A. Milne. 2015. Surface texture measurement for additive manufacturing.*Surface Topography: Metrology and Properties*, May 5.

Vandenbroucke, B., and J.P. Kruth. 2007. Selective laser melting of biocompatible metals for rapid manufacturing of medical parts.*Rapid Prototyping Journal* 13 (4): 196–203.

Zito, Damiano, Alessio Carlotto, Alessandro Loggi, Patrizio Sbornicchia, and Daniele Maggian. 2015. Definition and solidity of gold and platinum jewels produced using Selective Laser Melting technology. In*The Santa Fe Symposium on Jewelry Manufacturing Technology,* ed. Eddie Bell. Santa Fe, NM: Albuquerque, Met Chem Research.

Zito, Damiano, Alessio Carlotto, Alessandro Loggi, Patrizio Sbornicchia, Damiano Bruttomesso, and Stefano Rappo. 2014. Optimization of SLM technology main parameters in the production of gold and platinum jewelry. In*The Santa Fe Symposium on Jewelry Manufacturing Technology*. Santa Fe, NM: Eddie Bell (Albuquerque: Met-Chem Research).

Index